Dictionary of Aquaculture

Akhilesh Kumar Singh
Anil Mittal

2024

Daya Publishing House®
A Division of
Astral International Pvt. Ltd.
New Delhi – 110 002

First Published, 1999
Reprinted, 2024

ISBN: 9788170352037

Published by : **Daya Publishing House®**
A Division of
Astral International Pvt. Ltd.
– ISO 9001:2015 Certified Company –
4736/23, Ansari Road, Darya Ganj
New Delhi-110 002
Ph. 011-43549197, 23278134
E-mail: info@astralint.com
Website: www.astralint.com

Printed at : **Sagar Color**

FOREWORD

Indian aquaculture has evolved itself from the stage of a domestic activity to that of an industry through the years. With vast resources to the tune of about 2.3 million ha of ponds and tanks, 1.3 million ha of coastal waters, 3 million ha reservoirs, 1.3 million ha of beels, oxbow lakes, swamps that are amenable for fish culture, the practices have gained a great momentum during the last two decades. Recording an annual growth rate of 6% aquaculture sector is presently contributing to 1/3rd of the total fish production of 4.95 million tonnes in the country. The sector, apart from supplementing to the nutrition of masses, provides for rural development and employment in a big way as also high potentials for utilization and treatment of organic wastes, contributing to environmental upkeep. The potentials of the sector are estimated at 4.5 million tonnes that could be realised with adoption of modern technologies in different resources across the country.

At a time when aquaculture is making rapid strides, it is encompassing a host of disciplines like fisheries science, environmental management, fish genetics nutrition, pathology, engineering, economics, statistics, etc. It has become necessary to compile the various terminologies used in different discipline for quick comprehension and interactions among researchers and development officers involved in various stages of aquaculture development. It is heartening to note that Dr. Akhilesh Kumar Singh and Mr. Anil Mittal have made eftorts to bring out an Aquaculture Dictionary, that could be of great value for researchers, teachers, students, development officers of aquaculture. I compliment the authors for their efforts and hope that the book would be welcomed by one and all in the aquaculture sector as a useful addition to the information base.

S. Ayyappan

Director

CIFA, Kausalyaganga

Bhubaneswar-751002

FOREWORD

The Meaning of Aquaculture is similar to what agriculture connotes, the difference being that while the latter is concerned with terestial crops, the former is related to farming in water.

Aquaculture is a multidisciplinary subject, since it includes the culture of both plant and animal. Therefore, all associated branches of science, in the field of agriculture and animal sciences are involved in the field of aquaculture. It is for this reasons that a large number of terms are used in this subject, hence the necessity of a dictionary on the terminology in Aquaculture.

The meticulous work done by Dr. Akhilesh Kumar Singh and Shri Anil Mittal in collecting, compiling and collating the terms, used in aquaculture, is praise-worthy. The dictionary will be useful not only for the professionals in the field of aquaculture but will be a source of authentic terminology for students of various disciplines, especially those studying the 'culture'-agri and animal-*in water*.

M.A. Mohsin
Vice Chancellor
Birsa Agricultural University
Ranchi-834006

PREFACE

The Dictionary of Aquaculture is an effort to keep the pace with the rapid development and expanding vocabulary of the subject. Aquaculture has made tremendous development in last few decades. It is a multidisciplinary subject which includes Agriculture, Veterinary sciences, Soil sciences, Economics, Chemistry, Engineering etc. A need of Dictionary on the subject was realised long back when I was student and later on when teaching Fisheries to Veterinary Science under-graduates.

This Dictionary has been designed by keeping the needs of students and researchers of the subject. It will be also useful for teachers, scientists, development officers of the state departments, and many others interested in the subject. It includes the most frequently used terms which has been selected from different sub-group of fishery science, including Aquaculture, Craft and Gear, Fish Pathology, Microbiology etc.

An attempt has been made to write the terms in a clear and lucid style understandable even to non-specialists but consistent of the term.

While compiling the Dictionary it became necessary to draw upon the works of many authorities, I am highly indebted to them.

I am (A.K. Shingh) highly indebted to my family members wife Mrs. Bandana Singh, sons Alok and Anup Kumar for their co-operation in the compilation of the book.

At last but not the least we are thankful to all those persons who directly or indirectly helped us to complete the book.

There is more scope for improvement in this book, constructive suggestions for the improvement (if any) will be warmly welcomed.

A.K. Singh
Anil Mittal

A

ABC Acceptable Biological Catch subjectively estimated amount of catch of given species from a given region. Has no objective scientific definition.

Abiotic Factors Physical factors which affects the development/survival of an organism. e.g., water and minerals

Ablation Surgical removal of glands (eye stalk) to stimulate the maturation or growth.

Abrasion Resistance This is the resistance of a material or a specimen to wear of the whole or any part of it when rubbed against a surface.

Abscess A localised collection of pus in a cavity formed by disintegration of tissues.

Absolute Humidity It is the maximum amount of water vapour that a unit volume of air can obtain at saturation.

Absorption of Light When light penetrates into water it is gradually absorbed, that is transformed into heat. The rate of absorption, which is not the same for all wave lengths, is measured as the difference between extinction and scattering.

Abyssal Environment The environment at the bottom of deep ocean, one of the most constant bimoass. The temperature is between 0 and 2°C. The salinity is relatively constant (35‰) there is sufficient oxygen to support life. The life forms are suspension feeder, scavenger, carnivore, but their greatest difference from animals found at higher levels are their ability to live at constant high pressure (200 to 700 atm).

Abyssal Floor/Abyssal Plain It is the flat relatively featureless bottom of deep ocean. The average depth of the abyssal floor is about 4000 m. There are very few animal species that live in the inhospitable cold dark region of the ocean bottom.

Abyssopelagic Referring to water column, below 4000 m. The term refers to any organism or condition found in these regions.

Acanthella An acanthocephalan larvae lacks a proboscis and develops from a acanthor in the intermediate host.

Acanthocephala Small (about 1 cm) spinyheaded sausage shaped worms, all of which are parasitic and distinctly related to nematodes. These animals are named for the impressive hook, covered proboscis that attaches them to the gut of their host.

Acanthor An acanthocephalan larva, that develops from zygote and is infective to the intermediate host.

Accersory filaments Fibrillar structures such as striated lamella that may form part of the support in cells ,accessory filaments are distinct from microtubules.

Accessory Breathing An additional method of breathing besides the intake of oxygen through the gills which in some cases makes the gills dispensable. As substitute organs there is usually either a sac through the walls of which a great deal of blood flows or sac like folds in addition to the gills or the intestine or special lung like bags. Accessory breathing has in many cases been developed in such fishes that inhabit waters which are poor in oxygen content or have a muddy bottom. Fishes which possess accessory breathing organs must as a rule make use of their gills if the oxygen content of the water is rich enough.

Accident An unplanned event unexpected and undesigned, which occurs suddenly and at a definite place.

Accidental Species A species occurring in less than one fourth of a stand.

Acclimatization/Acclimation Gradual exposure to new environmental conditions in order to minimize shock or stress.

Acid a solution that contains a larger number of hydrogen ions.

Acid Cured Fish Fish and seafoods preserved in acidified brine or jelly with or without spices or other flavouring agents, in certain instances, salt is not essential.

Acid Death Point Death of fish due to the acidity of water. The pH 4.0 is considered lethal for fish.

Acid Sulfate Disease Syndrome in Prawn Normal metabolism of *P. monodon, P. indicus* is hindered resulting in retarded growth and eventual death due to low water and soil pH, yellow to orange to brown discolouration of gill and appendage surface is noticed.

Acid Sulphate Soil Acidic soils typically found where areas of mangrove have been cleared. Generally unsuitable for pond construction.

Acidification A process in which the pH falls or acid is added to decrease the pH. In anaerobic digestion problems may arise due to acidification following shock. Acid producing bacteria are capable of faster growth and metabolism than are methanogenic bacteria. Which are also less tolerent of acid condition. Hence the rate of acid production increases by any reason, the imbalance is enhanced as the acid producers grows faster and the acid utilising bacteria inhibited. Thus can lead to collaps of the biogas producing system.

Acidity / Basicity A measure of concentration of acid (base) in water or in a solution. The acidity or basicity of sea water is described by the pH scale. Pure water has a pH of 7, strong acid would have a pH of 1 or 2 and a strong base would have a pH of 12 and 13.

Acidosis When the pH of water goes down drastically owing to reduction of calcium salts or humic acid release is acidosis.

Aclid organ A structure with large tissue cutting spines which is located anteriorly on the acanthor.

Acquired immunity A series of host defences against a parasite which are characterized by extreme specificity and immunological memory, the defences are mediated by antibody and/or T cells.

Acrodont In some bony fishes, the teeth are more firmly attached, being fused with the underlying jaw bone. Such a condition is called acrodonty.

Acre Foot The volume of any substance which requires to cover one acre of a surface to a depth of one foot, equal to 43560 cubic feet (1232.75 cu.m.)

Acron Worm A type of hemichordates whose anterior (head) end resembles an acorn.

Act of God An event arising out of natural causes with no human intervention which could not have been prevented by reasonable care or foresight e.g. floods, earthquakes, wind storms.

Activated Charcoal, Carbon Finely divided carbon material capable of absorbing organic molecules.

Activated Sludge Material precipitated by removing organic material from sewage by saturating it with air and adding biological activated sludge.

Active capture gear Gear that is pulled through the water to activity entrap, entangle or hook fishes.

Active Storage Capacity The total amount of usable reservoir capacity available for seasonal or cyclic water storage. It is gross reservoir capacity minus inactive storage capacity.

Active Transport A biological mechanism that results in the transport of the molecules across a membrane against a concentration gradient. Active transport requires metabolic energy (usually ATP or an electro chemical gradient) and may be associated with a specific carrier protein or lipoprotein molecule.

Actuarial Describes the calculations made by an actuary. Essentially this is a branch of statistics dealing with the probabilities of an event occurring. Actuarial calculations if they are to be at all accurate require basic data over a sufficient period to permit likelihood of future events to be predicted with a certain degree of certainty.

Ad-Libitum Feeding until individuals or populations seem satisfied.

Adaptation A feature of an organism that have evolved under natural selection as it enables the organism to function efficiently in a specific aspect of its particular environmental niche. The adaptation can be either psychological or genetical.

Adaptive Disease The physiologic changes experienced by an organism because of exposure to a new environment.

ADC Apparent digestibility coefficient, during the passage of food through the digestive tract all of it is not digested and absorbed. The undigested portion is egested as faeces. The absorbed portion is determined by difference between the ingested and egested nutrients and usually expressed as a percentage of the amount ingested as ADC

$$ADC = \frac{\text{nutrient ingested} - \text{nutrient egested}}{\text{nutrient ingested}}$$

Adenoma A tumbour which consists of glandular tissues.

Adhesion A Pathological condition which results in connective tissue proliferation within and ground an organ.

Adhesive Eggs These eggs have an adhesive layer on their shell which becomes activated when the eggs comes into contact with the water. The adhesive layer sticks the eggs to some objects or to each other. Two types of attachment can be distinguished : (a) eggs sticking to objects. (b) eggs sticking only to each other, forming a clump or egg mass.

Adipose fin It is a fin made up of fatty tissue instead of the usual fin structure of spines and/or rays. It is behind, and separated from, the anterior dorsal fin. some times removed to mark fish because it does not regenerate.

ADP (adenosine diphosphate) A nucleotide consisting of adenine and ribose sugar with two phosphate groups attached.

Adsorption The adhesion of particles of one substance to the surface of another; such as the attraction of liquids of gases to a solid surface. In soils this includes the concentration of base salts (e.g. calcium and potassium) of microscopic size (colloids) surrounding larger mineral particles. Adsorption also refers to the process by which a thin film of water surrounds individual articles in sediments.

Adventitious Plant Parts Roots, spourts, buds, etc. which develop from the fully developed plant and not for vegetation points.

Aeration The mechanical mixing of air and water, generally to increase oxygen content but also to remove excess carbon dioxide. There are different types of aerator :

Total amount of oxygen added to 0.57 Hectare ponds during 4 hour periods

Types of aeration	*Oxygen added (kg)*	*Relative efficiency (%)*
Paddle wheel aerator	31.1	100
Crisafulli pump with sprayer	19.9	64
Crisafulli pump with discharging water from adjacent pond	12.1	39
5 HP spray type surface aerator	9.7	31
Crisafulli pump circulating pond water	7.5	24
3 HP spray type surface aerator	7.2	23
0.33 HP spray type surface aerator	2.5	8

Aerbiosis Biological processes that need oxygen.

Aerial Leaves Leaves of aquatic plants that rise above the surface of the water.

Aerobic Living or active only in the presence of oxygen.

Aerobic respiration Respiration in which free oxygen is used to oxidize organic substrates to carbondioxide and water with a high yield of energy. The overall reaction is $C_6H_{12}O6 + 6O_2 = 6CO_2 + 6\ H_2O$ + energy

It occurs in a number of stages the first of which (glycolysis) also occurs in anaerobic respiration in the cell Cytoplasm. With glucose as the substrate, a sequence of reactions results in the formation of pyruvate. The remaining stages which donot occur in anaerobic respiration take place in the mitochondria. Pyruvate is converted to acetyl coenzyme which enters a cyclic series of reactions. The Kirebs cycle with the production of carbondioxide and hydrogen atoms, produced at earlier stages are now passed to the electron transport chain. Here they combine with atoms of free oxygen to form water. Energy released at each stage of the chain is used to form ATP during a coupling process. There is a net production of 36 ATP per molecule of glucose during aerobic respiration and therefore the preferred mechanism of the majority of organism.

Aerobic Waste Treatment A biological process in which aerobic micro organisms are grown on waste water as a means of lowering the biochemical oxygen demand. Rapidly growing organisms use carbon compounds as a source of material for cell growth and metabolic energy generating cell biomass and carbon dioxide.

Aestivation The dormancy of certain animals during the dry season, the summer of a prolonged drought.

Aetiology The science of the cause or origin of disease.

Aflatoxins A group of highly toxic and often carcinogenic substances produced by the moulds *Aspergillus flavus* and *A. parasiticus* during their growth on food stuffs or animals feeds. There is a potential hazard to health from consuming even a very small amount of these toxins. The toxigenic Aspergillus species are widely distributed in nature and the liklihood of contamination is high especially in grains and oil seeds residues that are poorly dried or stored in damp conditions. More than a dozen forms of aflatoxins have been identified. Chemically the aflatoxins are substituted coumarins, B-1, being the most toxic and carcinogenic.

Agar Colloidal extract from red algae particularly from *Chondrus, Gelidiüm, Giyartina,* and *Gracilaria* spp. a solution in hot water sets to firm jelly which is used as a base for culture media for growing bacteria and as a gelling agent in food manufacture also used in the textile and medical industries.

Agglutinate The aggregation of organisms or particles into clumps, this may be due to the reaction of specific antibodies against surface antigenic, determinants on the organism.

Agglutinin Any substance which is capable of forming brides between antigenic determinants on contiguous cells to form visible clumps.

Agnatha Agnathans are vertebrates without jaws and paired appendages. Agnatha are the first vertebrates that appeared in the silurian period, about 390 million years ago. Early agnahans are known only as fossils. Lampreys and hagfishes are the only living agnathans. Modern agnathans are a menace to fishing industry in some parts of the world such as in the Great lakes of America, because of their parasitic habit.

Agricultural Wastes Products of agriculture that are not utilised by the established routes as food, animal feed, bedding, building materials etc. These wastes can be divided into three groups.—Woody or fibrous wastes such as straw bagasses, wood wastes etc. Manure and other products of animal husbandry, Solid waste from food processing industry.

AIR A mixture of gases predominantly nitrogen and oxygen. It also contains carbondioxide, water vapour and trace amounts of some other gases the constituents of air is :

Substances	*Percentage by volume*
Nitrogen	78.11
Oxygen	20.95
Water vapour	0.7
Argon	0.93
Carbon dioxide	0.01-0.1
Ozone	0-0.1
Neon	1.8×10^{-3}
Sulphur dioxide	$0\text{-}1 \times 10^{-4}$
Krypton	1.1×10^{-4}
Methane	2×10^{-4}
Xenon	5×10^{-5}
Nitrogen oxides	$0\text{-}3 \times 10^{-5}$
Hydrogen	5×10^{-5}

Air Breathing Fishes A fish capable of directly utilizing atmospheric oxygen for respiratory purposes, e.g. *Calarius batrachus, Heteropneusts* fossils.

Air Space of Soil In a representative silt soil, it has 30% of water space and 20% air space in pores.

Alae Lateral circular expansions along the body of some nematodes.

Alevin Early life stage of salmon and trout that possessing some traits of premetamorphosed larval fishes exhibits many morphological and behavioural features that are characteristics of juvenile fishes.

Algae An ancient group of primitive plants. Algae range from unicellular organisms to large, complex colonies of kelp and form blue green algae to yellow, green brown and red organisms. All species are autotrophic, some algae such as dinoflagellates and englenoids, have flagellae and motile.

Algal bloom A preponderance of algae in water which is turned completely green.

Algal Crashes An algal crash is the sudden death of a bloom and the release from this algal biomass of considerable amount of organic matter. These result stressed conditions for the pond. The concentration of carbondioxide and ammonium increase and there can be transformation of ammonium to toxic ammonia which can lead to fish mortalities. The concentration of nitrogen and phosphorus increases strongly owing to decay of organic matter settled on the sediments (ammonification processes). These decomposition remove oxygen from the water leading to anoxy and release of phosphate from the sediments. Several causes has been considered for algal crashes, Photolysis, attack by Pathogenic microorganisms nutrient deficiencies and grazing by microzooplanktons. The more probable reason for the death of algae is light and nutrient deficiency. Optimum fertilization will reduce the probability of an algal crash.

Algal scum Floating surface layer of algae in the fresh or, decaying condition.

Algal Toxciasis Toxins released by blue green like microcystis *Anabaena* and *Aphanizomenon* kill and other phytoplankton cause surfacing of fish stock. Persistance of the bloom will cause toxicosis for the fish stocks symptoms like convulsions leading to death.

Algicide Chemical that kills unicellular and macroalgae e.g. copper sulphate.

Alginic Acid Organic compound related to carbohydrates extracted from Brown algae particularly *Laminaria* spp used in the food industry as a stabilizer and a thickener its salts, the alginates are also used in industrial processes.

Alimentary Canal The passage in animals which extends from the mouth to the anus, through which foodstuffs pass to be broken down and absorbed into the body. The alimentary canal shows a great diversity of morphology and biochemical specialisation to suit a wide variety of diets.

Aliquot A single dose of a chemical or drug in solution.

Alkali Soil Soil with high degree of alkalinity (pH 8.5 or higher) or with a high exchangeable sodium content (15% or more of the exchangeable capacity or both.)

Alkali Water These water contain high level of alkali metal cations, usually sodium and potassium.

Alkaline Death Point Degree of alkalinity of water in which fish dies up, it is pH-11.

Alkalinity The concentration of basic minerals (e.g. carbonates) in the water, capable of neutralizing excess hydrogen ions (acidity). A number of bases including Carbonates, bicarbonates hydroxides, silicates, Phosphates, ammonia and various organic compounds occur in water. However HCO_3^-, CO_3^{2-} and OH^- are considered the predominant bases in natural waters. Since water with pH values above about 4.5 may contain bicarbonate water samples that turn yellow upon the addition of methyl orange indicator contain alkalinity. The alkalinity is expressed as mg/liter of $CaCO_3$. The standard sulfuric acid used for alkalinity titration is often of such strength that 1ml is exactly equal to 1mg of CaCo3. Alkalinity may be expressed as total alkalinity or as individual components of alkalinity i.e. hydroxide, carbonate, and bicarbonate alkalinity. In alkalinity estimation two indicators are used Phenophthalein and methyl orange indicator and titrated with sulfuric acid. The Phenophthalein alkalinity (P) and methyl orange alkalinity (M) togethr makes the total alkalinity (T)

(a) When T = P all the alkinity is due to hydroxide only

(b) OH^- and CO_3 does not exist together.

(c) When P is more than $\frac{1}{2}$ T but less than T, 2 M = 2 (T–P) = normal carbonate.

(d) Where P is less than $\frac{1}{2}$ T and M more than P

Normal carbonate = 2P

Bicarbonate = T – 2P

(e) When P = $\frac{1}{2}$ T or P = M normal carbonate = 2P = T

(f) When P = 0 the alkalinity is all bicarbonate = T

Alkaloids Complex basic organic compounds containing nitrogen, found in certain families of flowering plants. It differ widely in chemical structure and are classified on the basis of the type of heterocyclic group they contain, purine, pyrimidine, quinoline, isoquinoline, pyridine, reduced pyridine and indole. Most alkaloids are produced by plants and many have potent pharmacological activities. Alkaloids can act as analgesies, stimulants, muscle relaxants, opiates, vasoconstrictors, anaesthetics, tranquillizers, psychedelic agents.

Allochthonous Substances such as humus, silt, organic detritus plants and animal, produced outside and brought into the water body.

Aloe Leaf fibre obtained from the Aloe plants; a native of Brazil, but cultivated on commercial scale in Mauritius and India.

Alternative Technology Technology that aims to use resources sparingly and to do the minimum damage to the environment or to species inhabiting it which permitting the greatest possible degree of personal control over the technology on the part of its user.

Amarelo Cure Portuguese salt cod from which some of the salt has been removed between the stages of washing and drying by soaking the product in water; final salt content is about 18% and the product has a characteristic yellow appearance.

Amastigote The ovoid developmental stage (usually inter cellular) of a trypanosome in which there is no emergent flagellum.

Ambassis Genus of glass fish, worthless as food, family-centropomidae. The body is elongated to high dented at the forehead and more or less glassily transparent. The dorsal fin is in two parts. Another name for this genus is chanda.

Ambassislala / Indian glass fish The body is short and very much compressed, laterally. The forehead is deeply indented older specimens. The mouth is pointed and turned sharply upward, and the under jaw just forward slightly. D_1VII_3 $D_2 12$-15 A III/14-17, scales are very small. Av. size $2.^3/_4$ inches. Distributed in Northern India to Burma.

Ambergris A waxy substance found floating in the ocean. It is produced by sperm whales, and found in the digestive tract of these mammals, where it thought to protect the whale's digestive system against the hard beaks of squid, which forms the major part of the sperm whale's diet. When ambergris is fresh it is a black greasy, smelly substance. On drying it becomes grayish-black and smells faintly of musk. Ambergris was at one time used as a Pharmaceutical and a fixative, which is the part of a perfume that keeps the volatile aromatic materials dissolved in the perfume medium.

Ammocoete Premetamorphosis larval life stage of lamprays.

Ammonia The main nitrogenous excretion product of animals. Unionised (NH_3) fraction of ammonia is poisonous for fish. This portion increases with rising pH values and with temperature. Increasing the carbondioxide level decreases the toxicity of ammonia till the carbondioxide itself becomes poisonous for fish which reduces the pH value. The acute lethal concentration of ammonia for a variety of fish species lies in the range of 0.2 - 2.0 mg NH_3/l trout being the most sensitive and *Cyprinus* sp. being the most resistant.

Ammonification Ammonium salts are formed only when nitrogenous bodies are involved. They are the first really simple nitrogen compounds that appear as a process of microbial digestion goes on. Amino acids and similar nitrogenous materials readily yield ammonium compounds by enzymic hydrolysis. The transformation is ammonification.

Ammonifying Bacteria Those nitrifying bacteria (e.g. Pseudomonas) which break down the protein in the dead bodies of plants and animals to ammonia.

Ammonium Nitrate Used as a fertilizer and as a component of explosives, weedkillers and insecticides. The main source of nitrate run off from farm land resulting from excessive fertilizer use. In water it can be pollutant and may cause eutrophication.

Amphidromous Fishes which migrate to both marine and fresh water habitas, have wide range of Salinity tolerance e.g. gobies, chanos sp.

Anabaena It is a phytoplankton belongs to order-Pleurocapsdes, family; Nostocaceae. The genus is small filamentous, single, like nostoc but never colonial.

Anabas testudineus Climbing perch : A quite rugged and vigorous fish with rather parallel straight dorsal and ventral outlines. The mouth is terminal. The dorsal fin is as long as anal. D -XVI -XIX / 7-10, A IX-XI / 8-11. The scales are big. Av. length of fish upto 10 inches.

Anabolic Steriod Any steriod hormone or synthetic steriod that promotes growth and formation of new tissue. They are used in the treatment of diseases in agriculture or to boost livestock production and controversially in sport to build up athletes muscles. Most androgens have anabolic activity but when used by women they lead to masculinization.

Anabolism Metabolic reactions in which molecules are linked together to form more complex compounds. Thus anabolic reactions are concerned with building up structures, storage compounds and complex metabolities in the cell, starch, glycogen, fats, and proteins are all products of anabolic require energy provided by ATP produced by catabolism.

Anadromous Fish which lives most of its adult life in the sea but migrate to freshwater to spawn. e.g. Pacific Salmon.

Anaemia A deficiency in blood or red blood cells.

Anaerobe An organism that can live and grow in the absence of free oxygen. Anaerobe can be faculative in that they usually respire aerobically but can switch to anaerobic respiration when free oxygen is in short supply or obligate in that never respire aerobically and may even be poisoned by free oxygen.

Anaerobic Chemical processes occurring in the absence of oxygen. In ponds an indication of undesirable substrate conditions.

Anaerobic Biological Treatment Any treatment utilising anaerobic organisms to reduce organic matter in wastes.

Anaerobic Decomposition Organic breakdown in the absence of air commonly applied to the digesting of sewage producing methane.

Anaerobic respiration Respiration found in yeasts, bacteria and occasionally in muscle tissue in which oxygen is not involved. The organic substrate is not completely oxidized and the energy yield is low. In the absence of oxygen in animal muscle tissue glucose is degraded to pyruvate by glycolysis with the production of a small amount of energy and also lactic acid which may be oxidized later when oxygen becomes available. Fermentation is an example of an aerobie respiration in which certain yeasts produce. Only two molecules of ATP are produced by this process.

Anal Fin The unpaired fin that lies along the middle line on the ventral surface of the body behind the anus or vent.

Anatomy The science which studies the inner structure of live beings by dissecting them, also the inner structure of live beings in itself.

Anchor An implement for retaining a ship in position at sea by chaining it to the sea bottom. It is also used to secure fishing gear in a stationary position at sea. An outgrowth from the head of some organism it is used for attachment to the host.

Anchor Buoy Large spherical inflated buoy often covered with netting, supporting and marking the various ropes connected to the main anchor.

Anchorage A basis for attachment.

Anchoring disc A laminar Structure at the anterior end of a microsporidia spore.

Anemic ice fishes The anemic ice fisnes are the peculiar blood less fishes, belonging to the family chaenichthyid of the order perciformes. They are represented by about 16 species inhabiting the coolest water of Antarctica, the Kerguclen Bouvet and the South Georgia of the Atlantic Ocean. Anemic fishes are also known as ice fish and crocodile fish. These fishes contain only the white blood corpuscles while the red blood corpusceles or any other respiratory pigment are entirely absent.

Angle of Lay Angle of inclination formed by the strands with the axis of the rope.

Anglers People who catch fish for personal aesthetic experiences rather than solely for income or subsistence.

Animal Pole That part of the surface of animal eggs to which the nucleus (germinal vesicle) is closest. It is usually opposite the vegetal or yolky pole.

Animalcule A minute animal which cannot ordinarly be seen with the unaided eye.

Anion Exchange Capacity The sum total of exchangeable anions that a soil can absorb expressed as milliequivalents per 100 grams of soil.

Anonema The core of a cilium or flagellum which comprises the microtubule skeleton.

Anorexia The loss of appetite for food.

Anova Analysis of variance.

Anoxemia Pathological conditions resulting from lack of sufficent oxygen.

Antagonism Opposing effects which are produced by drugs, hormones etc; on living system. Adverse effect of one organism on the growth of another. The interferences may be due to the production of antibiotic substances or to competition for food etc.

Antenna The second and longer whip like head appendage of crustaceans used for sensing the environment.

Anterolateral Situated forward and on the side.

Anthopogenic Produced as a result of man's activities.

Anti Nutritional Factors It can be defind as those generated out of the natural food stuffs by the normal metabolism of the species involving different mechanisms and exerting effects inimical to optimal nutritional benefits.

Antibacterial Agent A compound or organisms that inhibits the growth of or kill bacteria.

Antibiosis A special case of antagonism in which an organism produces a substance injurious to other organisms.

Antibiotic Resistance The property of a micro organism that enables it to grow in the presence of a specific antibiotics. Antibiotic resistance is usually associated with a particular mutation or genetic characteristics which may be carried on a plasmid or other vector and transferred from one strain to another. Resistance is mesiated through action of various enzymes or changes in properties of cell membranes or ribosomal structure.

Antibiotic Sensitivity A measure of the susceptibility of a given microorganism to inhibition by a particular antibiotic.

Antibiotics A natural substance of relatively low molecular weight produced by a microorganisms which in dilute solution inhibits growth or destroys other organisms. Toxicity is generally selective. Most natural antibiotics whose structures vary widely are derived from the streptomycetes an exception being penicillin. The major antibiotics include compounds that block bacterial cell wall synthesis (penicillins and vancomycin) and protein synthesis (neomycin, kenamycin, streptomycin and the tetracylines). Antibiotics are used in man and other animals including aquatic animals as well as in microbiology and gene manipulation for a variety of purposes. Antibiotics are added to culture media in order to control contamination by foreign bacteria or are used to exert a selection pressure that permits the clonal growth of bacteria containing antibiotics resistance factor.

Antibody Protein produced by fish in an attempt to neutralise the effects of foreign material invading the body.

Antibody titre A measure of the amount of antibody usually in the blood of an animal.

Antigen A substance or cell that elicits a specific immune response when introduced into an animal.

Antioxidant Substance added to feeds to prevent or delay breakdown of fats (lipids) maintains food value of diet and increases shelf life.

Antiseptic A chemical compound that is used to destroy organisms that causes infection.

Apex Predators Top predator in an ecosystem, including mammals, birds, and man.

Apical At near or pertaining to the tip or apex.

Appendix A finger like posterior projections of the Ventriculus, common in some ascaridoid nematodes.

Application Factor The ratio between concentration of a substance that produce a particular chronic response in an organism and that which causes death in 50% of the population with a given period of time (the LD 50). The application factor is used to determine the maximum safe concentration of substances for the organisms under consideration.

Apron Old pieces of thick netting attached below the cod end as a chafing gear.

Aquaculture Farming in water, it is the aquatic equivalent of agriculture (the cultivation of terrestrial crops) and animal husbandry (the rearing of terrestrial animals). Aquaculture encompasses the culture of both animals (mainly finfish, crustaceans and molluscs) and plants (mainly sea weeds and higher plants) in both inland and coastal waters.

Aquaculture Insurance Similar to crop and livestock insurance generally against death or serious injury to the aquatic animals concerned is included in this class of business. The term does not generally used to describe insurance of farm buildings or equipments though the same insurer may cover both aquaculture production and fire insurance for farm buildings.

Aquafalfa Ground where the water table is high.

Aquanauts Scientists who explore the undersea world.

Aquaristics The study core and breeding of aquatic animals and plants in the aquarium done on a scientific basis, but without losing its esthetic and refreshing value for the biologically interested layman. Because of the fact that today aquaristics has spread to wide strata of the populations of all countries of the world, it is a good medium for diffusing biological knowledge in an easily understood form to the general public as well as to supply science with numerous explanation of the phenomena of life under water. Aquaristics plays an important role in this cooperation in the following fields, classification of the species of the native and foreign aquatic fauna and flora, breeding of animals and plants, fish diseases practical methods for keeping animals and plants ecology and ethology.

Aquarium Container made of glass or with glass walls which permits watching easily and for prolonged periods of time aquatic animals and plants, which inhabit it as well as their care and breeding. Aquarium can be of following types :

(i) According to the manner of construction, the all glass aquarium and frame aquarium.

(ii) according to the origin of animals and plants from, inland water or from the sea, the freshwater aquarium and the sea or salt water aquarium.

(iii) According to the temperature of water which is dependent upon the origin and vital demands of animals and plants the cold water and warm water aquarium.

Aquarium Cement A special black cement for filling out the putty joints in frame aquarium, which is used in liquid state, such cement must be guaranteed to be free of poison. Cementing should always be done after the aquarium has been in use for some time and after the putty has hardened sufficiently, as otherwise the cement will come loose easily due to the settling of the panes or else it is dissolved by the oil contents of the putty.

Aquifer A bed of rock that holds water and also allows water to flow freely through it.

Aquiprata Plants communities in wet meadows.

Argulusis/Fish Lice Argulus is a common parasitic copepods of fresh water fishes. It puncture the smooth portion of the fish skin, inject a cytolytic toxin through the oral sting and feed on blood. The feeding sites become ulcerated and hemorrhagic providing ready access to secondary infections by other parasites bacteria and fungi. Advanced infestations are characterised by erratic swimming especially flashing and loss of growth. The moribund fish appear exhausted and cannot maintain equilibrium particularly if the lateral line has been damaged.

Arrow Worm Tiny, transparent worm that lives near the ocean surface; active hunter with mouth bristles for hooking prey.

Artemia Small brine shrimp which lays drought resistant cysts. The cysts can be stored for several years and will hatch when placed in sea water to produce a nutritious nauplius larva, an ideal food for many crustacean larvae.

Artificial food Food not produced in the natural habitat, specially introduced food.

Artificial Impregnation Manual transfer of spermatophore from male to female placing near or inserting spermatophore into genital structures of the female so that sperm contact with spawned eggs is inevitable.

Artificial Seawater A solution of salts, resembling to a greater or lesser extent that of natural sea water.

Artificial Selection Selection of parental fish in a programme of breeding designed to produce offspring of progeny with specific characteristics.

Ascites The presence of fluid in the abdominal cavity.

Asian Fisheries Society (Indian Branch) Established in 1986. The permanent address is College of Fisheries, Mangalore. The Primary objective is to promote effective interaction and cooperation among the scientists and technicians involved in Fisheries research and development in India with a view to encouraging and facilitating research activity complementation, shareing of information and publication of research result. Create and propagate an awareness of the importance and the ways of sound utilisation, cultivation conservation and development of aquatic resources in the country.

ASP Annual surplus production An assumption in surplus production models that there is a biomass which can be removed by fisheries without changing population size.

Asphyxiation Suffocation due to dearth of oxygen.

Assembling the Net Joining the different sections of a net and combining it into one net and then hanging it on the head rope, foot rope etc. So that the net is ready for use.

Assimilation The process of in-corporation of simple molecules of food that has been digested and absorbed into living cells of an animal and conversion into the complex molecules making up the organism.

Atoll A ring of land surrounding a tropical lagoon and which in turn surrounded by a coral reef. An atoll is frequently formed of basaltic rock of volcanic origin and is capped with lime stone covered with coral. The islets that form atoll range in size from less than a mile in diameter to over 78 miles in diameter.

Atomisation The dividing of a liquid into extremely small particles so increasing it surface area.

ATP Adenosine triphosphate, the energy storage and transferring compound of most biological system.

Atrophy A decrease in the amount of tissue of the size of an organ after normal growth has been achieved.

Attachment plate A region of the trophont of a dinoflagellate which is modified for attachment of the host.

Aufeis Layer upon layer of frozen water discharged sequentially as by a spring.

Autogamy A kind of sexual reproduction in which the zygote is formed by fusion of two haploid nuclei from one individual.

Automatic Feeder Device powered by electric clockwork, water or air to feed fish at regular timed intervals.

Autospore One of the daughter cells formed by the internal division of one of the unicellular algae duplicating in miniature. The parent plant as in many of the chlorococeales.

Autothomous Turbidity creating matter produced within water body.

Autotrophic The quality which permits live beings to feed by transforming inorganic substances into organic ones, incorporating them in their body such as happens especially with green plants.

Autotrophic Organisms The organisms that can produce their own nutrients (food) such as Photo synthetic plants.

Auxotroph A strain of microorganism which has nutritional requirements additional to those of the organisms in its natural state.

Available Nutrient That portion of any element or compound in soil that can be readily absorbed and assimilated by growing plants.

Axenic Culture A pure culture one which has only a single kind of organism.

Axial structure A supporting structure aligned along the long axis of a cell.

Axostyle A rod like supporting organelle composed of microtubules arranged in sheets or ribbons, e.g. in the Trichomonadida.

Ayana Bark Bark of Ayana tree, widely used in India for tanning nets.

Azolla Azolla is a heterosporous fern belonging to the family Azollaceae (Salviniaceae) with seven living and twenty extinct species. Based on the morphology of reproductive organs the living species are grouped into two *A. fileuloides, A. microphylla A. rubra* and Rhizosperma (*A. pinnata, A. nilotica*). Proliferation of Azolla is basically through vegetative propagation, but sexual reproduction occurs during temporary adverse environmental condition with the production of both microsporocarp and megasporocarp. It is useful in Aquaculture practices—primarily as nitrogenous biofertilizer. The mean daily nitrogen fixing rates of a developed Azolla mat are in the range of 1.0-2.6 kg/ha.

B

Babul Bark Bark of the babul tree *Accacia arbabica* used as a source of tannin.

Back cross The crossing of a hybrid back to the original parent generation. If a homozygous dominant AA is crossed with a homozygous recessive aa the F2 generation obtained by selfing the F1 would be 25% AA, 50% Aa and 25% aa. To distinguish between the phenotypically identical AA and Aa these can be back crossed to the homozygous recessive parent as a test cross. The off spring from AA X aa will all be identical (Aa) whereas in the cross Aa x aa 50% will have the dominant phenotype (Aa) and 50% will show the recessive character (aa).

Back Waters Stagnant brackish water bodies that connected with the sea during rains while in other seasons become detached.

Backyard Hatchery Small family owned and run hatcheries.

Bacteria Single-celled organisms that lack a nuclear membrane.

Bacterial Gill Disease Caused by myxobacteria. Fish developing bacterial gill disease show sudden lack of appetite; orientation against the current and distribution of individuals equivalent from other. Gill lamellae show proliferation of epithelium which in advanced cases result in clubbing and fusing lamellae or even filaments.

Bacterial Kidney Disease The disease produces by corynebacterium. The main symptom is exophthalmus, large external welts. Tiny unbroken pimples are often associated with lesions. Kidney typically light and grossly swollen. Liver pale, Visceral cavity and other organs are typically hemorrhagic.

Bacteriocidal The term Used to describe a compound that has a lethal effect on bacteria. A bacteriocidal compound may act by interfering with a vital biochemical pathway or by destroying the molecular structure of the cell.

Bacteriostat Substances which retards or inhibits the growth of bacteria.

Bacterium Single celled microscopic organism with a rigid cell wall but no separate nucleus and no chlorophyll, about 1 micron in diameter.

Bagridae African and Asiatic family of cat fishes belonging to the sub-order of the siluroidea. Which possesses a moderately elongated body and a completely naked skin. The dorsal fin inserts relatively near the front end. An adipose fin is always present and nearly all species disproportionately big. The fish possesses 3-4 pairs of barbels, the jaws are always equipped with teeth. The big and independent swimming bladder is well developed.

Bait Any substance that by its appearance or smell is used to lure fish to take a hook or enter into a pot or trap.

Bait Fish Fish cultured for use as live bait by anglers.

Bal Bakwa Salted fish product, usually a whole large fish. Approximately 20% by weight of salt is added to allow controlled bacterial action during the six to eight months, ageing process, usually warmed in vinegar before serving. It is common in Phillippines.

Balance of Nature A state of equilibrium in nature resulting from the interaction of all elements of the environment.

Balance Sheet A financial statement that lists the firm's assets and liabilities (or claims against its assets)

Banana Link A metal link fitted to the apex of the brackets of an otter board to prevent the C-link assembly from being fouled.

Banded Barbs A group of small fishes belonging to the family of the carp like or cyprinidae genus Barbus. The several species are very similar and are only unreliably differentiated externally by the number and disposition of the black transverse bands. In the case of many banded barbs which today are still being considered as different species.

Bank Storage Water absorbed and stored in the soil cover in the bed or bank of a stream, lake or reservoir and returned in whole or in part as the level of the surface of the water body falls.

Banks Offshores elevations or shallow areas of the continental shelf that are separated from each other and the mainlanded by broad deep channels.

Bar (Millibar) Measure of pressure in the atmosphere and in the ocean. 1mb (millibar) = 0.750 mm Hg.

Bar Cut See cutting netting to shapes.

Barb Like Fishes A term for a group of small tropical and sub tropical carp like fishes which resemble the Barbs particularly the genera Rasbora, Damie, and Esomus.

Barbecued Fish Fish roasted or grilled over an open charcoal fire and served hot.

Barbels They are soft, thread-like filaments around the mouth, varying in number from 4 to 8, and sensitive to touch.

Barbus binotatus c.n. Booty Barb. The body is slender and elongated, the terminal mouth with four barbels. D-12 A-8. dis-In ponds rivers and lakes from Malayan Islands up to the Philippines, quite frequent everywhere.

Barbus chola The body is laterally compressed looks quite compact. Head & mouth are small; the mouth opening is slanted upwards and carries one pair of barbels. D-11, A-7 Av length 5 inches. dis-The species is quite frequent in all kinds of water even paddy field of East India and Burma.

***Barbus conchonius* (Ham) c.n.** Rosy barb. The body is long and slender. The species possess no barbels D-10, A 7-8 length upto 6 inches. Dis-In standing and slowly flowing waters of Northern India quite frequent every where.

***Barbus stigma* c.n.** Dotted barb. i.e. Elongated with slightly arched body compressed laterally. The mouth cleft is filted obliquely upwards. No barbels present. D-11-12, A-8, length 8 inches. dis-In waters of all kinds in India and Indo-China.

Barbus terio **(Ham)** Body elongated with dorsal and ventral contours equally arched, compressed laterally. The mouth cleft is slanted obliquely upwards. Barbels lacking D-11 A-7, length upto 3 inches. dis-In water of all kinds in Bengal also in paddy fields. A quite common fish.

Barbus ticto **(Ham) c.n.** two spot Barb. Body moderately elongated, dorsal and ventral contours equally arched, compressed laterally. The obliquely rising mouth cleft carries no barbels D-11, A-7 length upto 3 inches.

Barbus/puntius **(With Barbels)** A genus of small medium size or big fishes belonging to the family-cyprinidae.

Barnacle Sessile crustacean with overlapping calcium carbonate plates and cirri.

Baroclinic Conditions in the ocean and in the atmosphere where constant pressure surfaces intersect constant density surfaces.

Barrage Pond The pond created by construction of a dam/ dike across a gently sloping flat valley.

Barrier Spit A bench barrier connected to the main land at one end and terminating at the other end to open water. As its build down drift of the longshore currents it may develop a hook at open end as currents are refracted shorewards.

Base Exchange It is generally assumed that adsorbed nutrients are rather readily available both to higher plants and to micro organisms. As how is the transfer from soil colloid to soil organisms doubtedly facilitates the availability of nutrients held in this way.

Base Level Lowest level to which stream can erode its channel, sea level, lake level or the level of the main stream into which the tributary stream flows.

Base Lines The tolerance levels of organisms to a particular concentrations of substances. These vary from species to species.

Base Saturation Percentage The extent to which the adsorption complex of a soil is staturated with exchangeable cations other than hydrogen and aluminium. It is expressed as a percentage of the total cation exchange capacity.

Basement membrane The extracellular supporting layer of mucopolysaccharides and proteins underlying the epithelium.

Basket In which long line gear is arranged and kept. The term is used to indicate a unit of long line which is comprised of one main line and five or six or more branch lines with hook.

Bast Fibre obtained from the inner bark of plants such as jute, flax, etc.

Batch Culture A system of rearing prawns involving the total harvest by seining or draining or both, at a certain interval after stocking. The ponds are then drained before restocking.

Bathythermograph This is an instrument for measuring and recording the water temperature, versus depth on a metalized glass. The instrument has the depth range of 0-275m and a temperature range from -1 to 30°C with an accuracy of ± 0.2°C.

Bati A small cup like measure, usually of 130-150 cubic centimeter capacity used for measuring early carp fry.

Battery Culture Culture of crustaceans in individual compartments or multiple tank in a controlled indoor environment.

Beach Barrier A wide range of sand protruding above normal high tide level parallel to the coast and separated from it by a lagoon or marsh. There are two proposed modes of its origin (a) Barriers are formed by long shore drift (b) Barriers can be produced by wave action transporting sand shoreward and turbulence due to surf piling sand on to the barrier side. This is augmented by long shore drift.

Beam A metal or wooden beam to which the beam trawl net is attached during operation.

Beels Shallower and smaller lakes.

Bekko Tortoise shell used for ornaments such as necklaces, rings, pipes cases, combs etc.

Belly The central netting section of a trawl between square wings and throat or cod end.

(i) False Belly : Old pieces of thick webbing or raw hides, attached below the belly of a trawl as a chafing gear. It is also known as rubber.

(ii) Top belly : The portion of webbing that forms the lower body of the trawl from the foot rope to the fore part of the throat or cod end.

Belostoma/Giant Water Bug Insect, belong to order Hemiptera family Belostomatidae. The body is brown or dull greenish in colour, body large flat and oval, head bears a powerful beak, foreleg chelate and hindlegs long modified for swimming wings functional grow upto 20-70 mm.

Bench Mark A mark cut in stone that is used as a reference point in surveying to determine altitude.

Bends Gas bubble disease caused when blood becomes supersaturated with gases which then come out of solution in capillaries and damage tissues.

Benthic Organisms/Benthos Organisms living on or in the bottom sediments.

Bentonite A very fine grained clay with a high shrink/swell potential; often used to seal ponds.

Berm/Bench Line A platform like space between bund and watery area. The berm may vary between 0.5 and 1.0m depending on the height of the bund and the size of the pond. The berm apart from serving as a walkable space for fishermen also protects the bund from direct contact with water.

Berried Female crustacean carrying eggs under her abdomen during a period of incubation.

Betta splendens (Re) c.n. The Siamese Fighting Fish. Body not excessively elongated, DI/8-9, AII-IV/21-24, length upto 3 inches. Dis-Singapore, Malaya, Indo-China in standing and slowly flowing small waters with swampy bottom.

Bhasa-badha fishery Low lying marshy salt water areas under reclamation in the sunderbans subject to tidal effect, embanked and provided with sluices to trap in fish seed to grow.

Binary fission A sexual reproduction by an organism which results in two progenies.

Binder Natural or artificial substance added to bind and hold finely ground dietary ingredients together when in water.

Binomial nomenclature A system of classification introduced by linnaeus the swedish botanist in which each species is given two names. The first is the generic name written with a capital letter which designates the genus to which the species belongs. The second is the specific name indicating the species. The generic and specific name in latin and are printed in italic type. For example the Rohu is known as *Labeo rohita* the Labeo is the generic name and rohita is the specific name.

Binoro Small fish, mackerel, sardine or other fish brined drained and packed in dry salt, common in Philippines.

Biochemical Oxygen Demand (BOD) A measure of capacity of the organic materials in a solution (usually an effluent) to takeup oxygen and become oxidized. The amount of oxygen consumed by respiratory processes in a bottle of water incubated in the dark at constant temperature for a specific time. Standard biochemical oxygen demand determinations are for 5 day' incubation at 20°C.

Biocide Any agent that is able to kill living organisms. Some times used as synonym for pesticides.

Bioclimatology The science of the relationship of life to climate.

Biocoen All the living components of the environment.

Biodegradable An organic substance that is quickly broken down by normal environmental process.

Biofertilizer Fertilizers / Manures made by the action of living organism.

Biofilter/Biological Filter A water treatment system in which there is a large surface area occupied by micro-organisms that oxidize dissolved organic matter, ammonia and nitrate to less harmful products.

Biogenous Decalcification and Calcification Biogenous calcification and decalcification are parts of the circulation of substances in water which is extra ordinarily complex. Among water and all its components, climatic influences, the bottom the plants, the animals etc. exists an intimate mutual interaction. Under the influence of light in all plant stocked waters there is a rhythmic exchange of substances which is based on the assimilation of the plants. During this the free carbon dioxide dissolved in the water is taken in by the plants. After this is used up carbon is taken from bicarbonates which too are dissolved in water. Thus a simple carbonate of calcium is developed in the water precipitating and settling on plants the bottom etc., this process is called biogenous decalcification.

Biogeography The study of the geographical distribution of animals and plants. Marine biogeography includes description and distribution of Marine animals and plants and analysis of factors which determine the distribution and abundance of a given species.

Biological Bench Marking The use of plant or animal species to measure pollution based on assessment of population level and fitness against which changes can be evaluated.

Biological clock The internal mechanism of an organism that regulates circadian rhythms and various other periodic cycles.

Biological Monitoring The direct measurement of changes in the biological status of a habitat based on evaluations of the number and distribution of individuals or species before and after a change.

Biomass The total quantity of living organisms or specific organisms in a defined body of water.

Biome A major regional community of plants & animals with similar life forms and environmental conditions. It is the largest geographical biotic unit and is named after the dominant type of vegetation, such as tropical rain forest, grassland or coral reef.

Biophysical Pertaining to the living organisms and water conditions of the environment.

Biorhythm A periodic physiological or behavioural change that is controlled by a biological clock.

Biosphere The part of the earth and its atmosphere that is inhabited by living organisms. The earth's surface and the top layer of the hydrosphere (water layer) have the greatest density of living organisms. The geosphere or non-living world is made up of the lithosphere (solid earth) hydrosphere and atmosphere.

Biota Animals and plants of an ecosystem.

Biotechnology Use of living organisms in industrial production processes.

Biotic Factors Factors of a biological nature such as availability of food, competition between species, predator and prey relationship etc. which besides the purely environmental factors also affect the distribution and abundance of a given species of plant or animal.

Biramous appendage A two branched appendage found in crustaceans. It consists of three sections the protopodite (Basal section) which is attached at one end to the body and at other to the two branches the expodite (Outer branch) and endopodite (inner branch) Each section may consist of several segments.

Bismark Herring Herring block fillets or whole herring headed and gutted, cured in acidified brine usually in barrels or other containers, after finished curring packed with acidified brine of a lower vinegar and salt content, also with slices of onions, cucumbers carrots and spices also with sugar added.

Bisque A puree or thick, soup made from crustaceans. In France the basic material is lobster and crayfish.

Bivalve Molluscs with an openable shell oysters, clams, mussels etc.

Black Gill Disease in Prawn Prawns are affected due to chemical contamination like cadmium, copper, oil zinc, potassium permangnate, oxone, ammonia, and nitrate. Ascorbic acid deficiency, high siltation, high organic load, debris etc. contribute to the disease of the gill shows reddish, brownish, to black discolouration and atrophy at the top of filaments. In advance all the filaments becomes totally black, loss of apetite and leading to mortalities.

Black Mud A foul smelling sediment, rich in hydrogen sulphide and organic content typically occurring in poorly managed ponds and uncleaned tanks and pipe work.

Blanched Processing term meaning precooked or parboiled *e.g.* at 65°C for 15-20 seconds.

Bleaching Powder Chlorinated lime (i.e. Calcium hydroxide more or less saturated with chlorine) One of a number of chemical agents that will bleach or destroy natural colour, so rendering substances white but has largely been displaced by elemental chlorine and especially by sodium hypochlorite.

Blowhole In whales and dolphins, opening on top of head used for breathing.

Blubber Unrendered subcutaneous body fat of whales or other aquatic mammals.

Blue Clawed Male Large dominant male *Machrobrachium rosenbergii* that has developed blue chelae.

Blue Disease The disease with sky blue colour instead of normal brown black, absence of intense red colour after cooking is observed in intensive culture systems towards the end of grow out culture period. Lethargic shrimps with shells sometimes soft and than with rough surface are noticed in affected animal with disrupted tubules in the hepatopancreas. This disease is mainly due to acid sulfate soils, high organic waste, low dissolved oxygen levels.

Boat Rope In gill net operation the rope containing the craft to the bridles at the proximal end of the fleet.

Body Cramp in Prawn Juveniles and adults show partial or complete rigid flexure of tail possibly due to mineral imbalance or increased water and air temperature. Partially cramped shrimps swim with humped abdomen, where as fully cramped individuals lie on their sides at the pond bottom, resulting cannibalism of cramped shrimps by unaffected ones or death.

Bohr Effect Carbondioxide affects fish respiration. High concentration of carbondioxide interfere with the capacity of hemoglobin to fix and transport oxygen and even cause release of oxygen already fixed. This phenomenon is known as Bohr effect and it slows down the metabolism of fish.

Bomb Calorimeter A sealed insulated container used for measuring energy released during combustion of substances (e.g. foods and fuels).

Boned Fish Fish from which the principal bones have been removed.

Boneless Fish Fish containing no bones at all.

Bony Fishes Have bone skelton, loose scales on skin, swim bladder; e.g. tuna.

Borehole Shaft drilled down to provide water supplies from natural aquifers or ground water.

Boring tooth A cuticular tooth like projection at the cephalic end of a larval nematode.

Bottom Dredge (Biological) This is a compact sampler to be towed along the bottom for collection of biological or mineral specimens. The dredge has a strong stainless steel rectangular frame with a size 50×20 cm (approx). It is delivered with a pair of folding varms and a bridle for connection to the towing line.

A sturdy net bag of 6 mm mesh size is attached to the rear of the frame and this is protected by a heavy canvas cover. The net bag is 6 cm long.

Bottom Drift/Bottom Set The method of operation of the gill nets when it is allowed to drift close to the bottom layers of water.

Bottom feeder A fish that generally feeds at the lower levels of water immediately above the bottom.

Bottom fishes Fishes that reside at the bottom of the water column spending much of their time at or near the surface of the substrate.

Bottom Knot Knot at the bottom of a mesh.

Bottom Sampler (Lenz) This the modified design of Ekman brige. To obtain a review of the relief surface the sample can be separated by the dividing sheets into 5 layers of 20 mm thickness. These layers can then be removed separately for examination. The operation is same as bottom sampler Ekman brige.

Bottom Sampler (Van Veen) The sampler can samples from any desired depth. The effective surface of the bottom sampler when operated is 250 cm^2. While it is being lowered both levers are locked wide apart whereby the jaws are opened. Upon making contact with the water bed the locking mechanism is released and when the rope is pulled out to raise the sampler the jaw close.

Bottom Sampler/Ekman Brige This instrument consists of a sturdy walled brass cabinet with an effective grasping area of 225 cm^2. The bottom of the sampler body consists of 2 spring closed jaws, the mechanism of which is released by a messenger. Two light plates over the opening at the top of the sampler prevent the sample from being washed out while the sampler is being heaved out of the water.

Bottom Trawl A conical shaped net towed along the bottom.

Bow Line A knot commonly used for making a non slip loop on the end of the rope.

Bow Line Knot See knot

Boxed Stowage Stowage at sea in boxes. The ice is placed above and below the fishes.

Brace Line The lines used for lacing the adjoining shots in a fleet of gill nets.

Brachiopoda A small phylum of marine invertebrates, the lamp shells (e.g. Lingula) living attached to a firm substratum in shallow water. Lamp shells superficially resemble bivalve molluscs in the possession of a bivalve shell but the valves are placed dorsally and ventrally (as opposed to laterally; as in bivalves). The internal lophophore a filter feeding organ composed of cilliated tentacles protrudes from the shell. Excretion is carried out by nephridia. The sexes are separated and there is a free swimming larva. The phylum was very large in plaeozoic times.

Brackish Marine and esturaine waters with salinity between 0.5 and 30 ppt. It is due to the mixing of seawater and freshwater.

Brackish Water Sea water diluted with fresh water for example an estuary.

Braided Rope Rope in which the strands are not being tied together by twisting but by interlacing in such a way that they cross each other in a diagonal pattern.

(i) Solid braided rope : Braided rope in which all stands run in the same direction either towards left or towards right.

(ii) Tubular braided rope : Rope braided in the form of a tube and is the most common.

Branch Line Proper Is the main part of branch line of tuna long line and measures about 12 metres in length. On one end it is connected to main line and the other end is attached to the Swivel.

Branchia A Leaf like or finger like lateral appendages on the urosome of some leeches.

Branchial glands These are specialized cells of the epithelium. The most common specialized branchial glands are the mucous glands and chloride cells.

Branchiostegal rays Internally the operculum is lined by a flexible branchiostegal fold supported by bony sabre-shaped branchiostegal rays attached along the ventral border of epi-and ceratohyales and another unpaired basibranchiostegal bone lying posterior to the basihyal. The branchiostegal fold serves one-way valve permitting the water only to leave the gill chamber through the opercular opening.

BRD (By catch reduction device) Mechanism placed in shrimp trawls to reduce the incidental capture of fishes particularly juvenile life stages.

Break Even Point It can be defined as the amount of income where the income (minus variable cost) is sufficient to cover the fixed costs and there will be no profit and no loss. The break even price then be the total operating cost divided by the quantity of production and the break even production, the total operating cost divided by the unit price of the product.

Break Water A barrier built into the sea to absorb the impact of waves and protect the coast.

Breaking Load/Breaking Strength/Breaking Force Minimum load required to rupture a specimen length of fibre or yarn.

Breeder/ Brood Fish Sexually mature fish ready to breed.

Breeding Cycle A period between hatching and the first spawning of a given generation.

Breeding Hapa It is a box like rectangular container provided with ropes at its four top and bottom corners for tying with pole. All the sides of the hapa are stitched and closed except the top portion which have arrangements (jip or hooks) to close or open as per need. The material is fine meshed cotton cloths or nylon. Various size of breeding hapa can be made according to size groups of fishes *e.g.* 3.5m (length) × 1.5m (width) × 1.0m (depth) is usually used for over 5 kg fishes.

Bridge Stone Cage Recently developed sea cage or pen which has a flexible float or collar, because of flexibility it can be used in exposed environments.

Brine Solution of salt in water.

Brine Shrimp Common name for *Artemia* used as live food for fish larvae.

Brined Fish Fish that have been immersed in brine as a pretreatment to further processing.

Broad-cast spawning Release of eggs and sperm into the water column, once fertilized, buoyant eggs of broadcast spawning fishes drift with currents while suspended in the water column.

Brood Stock Pond The ponds constructed for brood fish.

Brook A small stream or rivulet.

Broth A culture medium derived from a protein solution or hydrolyzate more generally a suspension of cultured microorganisms.

Buccal ciliary apparatus The specialized ciliature inside and around the buccal cavity in ciliates for food uptake.

Buccal Pertaining to the oral cavity.

Buckling Large fat herring, sometimes headed or nobbed, lightly salted and hot smoked, also called pickling.

Buffer A solution that resists any change in acidity or alkalinity. Buffers are important in living organisms because they guard against sudden change in pH. They involve a chemical equillibrium between a weak acid and its salt or a weak base.

Bulk stowage Stowage at sea of fish mixed with ice in layers about 4.5 cm deep, each layer additionally protected by several cm of ice top and bottom.

Bull Rope See lazy line.

Bund Bund are the protecting structures of fish ponds and are essential for drainable and undrainable fish ponds.

Bundh It is a dike or an embankment. In regard to fish breeding, bundhs are special types of perennial or seasonal ponds or impoundments where riverine conditions are simulated during the monsoon season owing to the accumulation of rain water from catchment areas (also known as acude in Brazil) See wet bundh and dry bundh.

Bunt In the end part of purse seine net which is shot and hauled up when the catch is accumulated for brailing and is made of stronger webbing.

Buoy A float with visible mark and sometimes carrying a flag or lamp attached to a fishing gear to indicate the presence of the gear or to help locating the gear.

Buoy Rope Rope connecting buoy to that part of the gear being supported or marked.

Buoyancy Capacity for floating lightly on water. (i) Extra Buoyancy : The net upward thrust of the float when fully submerged in water.

Business Risks Risks directly related to the business of producing aquatic animals and plants. It includes production, market related and consumer related risks.

Buster Colloquial term for a crab that has just started to moult.

Butter Fly It is an angle bar of flat iron, one side of the angle being shorter than the other. The shorter side is connected to the foot rope and the larger side is connected to the head rope. Its purpose is to spread the wings vertically.

Butterfly Shrimp A form of prepared value added shrimp.

By catch-incidental catch Organisms that are captured incidently during fishing activities directed toward other texa.

By Product A substance obtained during the manufacture of a main product.

C

Cable Laid Rope A 9 strand rope in which 3 primary ropes are twisted together without a central rope.

Cachexia The wasting of body tissues due to extreme malnutrition.

Caecum An anterior projections of the intestine from the region of the oesophageal intestine junction common in some ascaridoids and spirurids.

Cage Culture Raising fish from the juvenile stage to commercial size in a volume of water enclosed on all sides, including the bottom, while permitting the free circulations of water through the cage.

Calcareous Made of or containing calcium carbonate.

Calcification The deposition of calcium compounds occasionally associated with capsule formation.

Calcimorphic Soil Soils containing excess limey material.

Calcium Hardness The calcium concentration in water is normally expressed as calcium hardness in terms of equivalent $CaCO_3$.

Calcium Hydroxide/Slaked Lime It is produced by the action of water on calcium oxide. Used in sugar purification, glass production, agricultural, etc.

Calibration Measuring and adjusting the performance of a system to a specified standard.

Calipash The fatty greenish flesh from the carapace of the green turtle. The meat is also dried and more rarely smoked.

Calipee It is a fatty gelatinous light yellow substance obtained from the plastron of turtles; esteemed as delicacy.

Calorimetry The measurement of thermal changes involved in chemical or physical reactions, in either in vitro system or an intact organism. For example a bomb calorimeter is used to determine the calorific value of foods. This is a steel chamber in which the food stuff is placed and ignited in the presence of oxygen. The increase in temperature of water in a jacket surrounding the calorimeter indicated the heat produced by the oxidized food.

Cambered Otter Board Otter board that is curved in fore and aft direction.

Canned Fish Fish packed in containers which have been hermetically sealed and sufficiently heated to destroy or inactive all micro-organisms that will grow at any temperature at which the product is likely to be held and that will cause spoilage or that might be harmful.

Cannibalism Consumption of one individual by others of the same species.

Canthaxanthin Pigment often used in diets to colour fish flesh.

Canthocamptus Belongs to Arthropoda group, order copepoda family Harpacticidae. Body linear, Cylindrical, Completely segmented last thoracic segment included in the Urosome, egg sac single cephalothorax not sharply differentiated, antennae short with 8 segments.

Carapace The one piece shell structure covering head and thorax of crustaceans.

Carbohydrate A class of compounds occurring widely in nature and having the general formula type Cx (H_2O) Y. Carbohydrates are generally divided into two main classes sugars and Polysaccharides. Carbohydrates are both stores of energy and structural elements in living systems. plants having typically 15% carbohydrate and animals about 1% carbohydrate. The body is able to build up Polysaccharides for simple units or break the larger units down to more simple units for releasing energy. Carbohydrates requires neutral or slightly alkaline conditions for the operation of enzymes such as maltase and amylase.

Carbon An essential element in plant and animal nutrition that occurs the all organic compounds and thus forms the basis of all living matter.

Carbon Cycle The circulation of carbon between living organisms and the environment. The carbondioxide in the atmosphere is taken up by autotrophic organism (mainly gren plants) and incorporated into carbohydrates. The carbohydrates so produced are the food source of the heterotrophs (mainly animals). All organisms return carbondioxide to the air as a product of respiration and of decay. The burning fossil fuels also releases Carbondioxide. In water carbon combined as carbonates and bicarbonates, is the source for Photosynthesis.

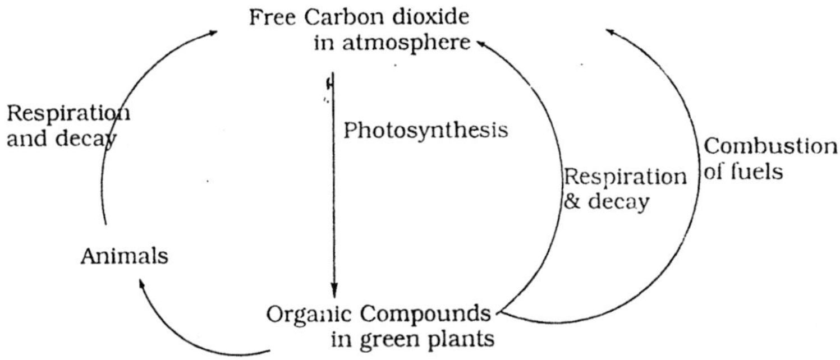

Carbon Monoxide A poisonous gas with no colour and little smell, formed when combustion occurs without much air.

Carbon Nitrogen Ratio (C:N) The ratio of the weight of organic carbon (C) to the weight of total nitrogen (N) in a soil or in organic material.

Carboxy Haemoglobin The result of the combination of carbon monoxide with blood haemoglobin. While haemoglobin has an affinity for oxygen, its affinity for carbonmonoxide is greater.

Carcinoma An epithelial tumour which tends to invade the lymph spaces of surrounding connective tissues

Caridea Taxonomic group (infra order) of shrimp and prawn.

Carnivorous Plants Plants which have developed the ability to trap insects and other small animals, dissolving them and feeding on their proteins by means of digestive process.

Carp A group of freshwater fishes with toothless mouth.

Carrageenin Water soluble edible colloidal extract from Red Algae.

Carrying Capacity The population of a given species that an area or volume will support without undergoing deterioration. When standing crop reaches a level at which the natural food is sufficient only for maintenance and no food is left for growth, growth ceases entirely. This will be the carrying capacity of pond for that species.

Cascade A repetitive system for separation of components as the process is repeated The principle is used in many industrial processes including the separation of nuclear isotopes including uranium.

Cat's Clay Soil developed from marine sediments containing deposits of sulfide compounds. Cat's clay is often found along coastal plains and pond constructed in this land may have extremely acidic water.

Catabolism Destructive metabolism, destruction of living matter.

Catadromous Fish that lives most of its adult life in freshwater but migrates to seawater to spawn. e.g. eels.

Catamaran This is made up of logs tied together. They can be used both in inland waters and along the surf beaten coasts without a shelter.

Catarrhal The discharge of fluid due to inflammation of the mucous membrane.

Catchment From where a single river system collects its waters. The boundary of a catchment area is defined by the watershed.

Catfish A group of scaleless fishes, often with long whisker like feelers.

Catheter It is a L-shaped fine polythene tube or glass-tube, which is inserted into the oviduct throught the vent of the female to draw a small quantity of eggs for microscopic examination.

Cation Exchange Capacity The sum total of exchangeable cations that a soil can adsorb sometimes called total-exchange capacity. It is expressed in milli equivalent per 100 grams of soil.

Cation Exchange The interchange between a cation in solution and another cation on the surface of any surface active material such as clay or organic matter.

Catla Sc. n. *Catla Catla* (Ham) fam. c.n. Katla, Bhakura, dis. Throughout India upto Krishna river, Pakistan, Nepal, Bangladesh, Thailand, i.c. Head broad, Body short deep, upperlip absent, lowerlip moderately thick, lower Jaw with a movable articulation at the symphysis but without a prominent Knob, Dorsalfin long, inserted above tip of pectoral fin with 17-19 rays and without spines, Anal fin short with eight rays caudal forked, scales moderate, lateral line complete with 40-43 scales.

Cattle Manure The cattle are ruminants and the food stuffs are repeatedly masticated making the excrement quite delicate. Cattle manure contains less nitrogen than Pig manure (C : N = 25:1) cattle urine contains hippuric acid so it decomposes slowly the average daily excreta is 25 kg/cow in which the faeces to urine is about 3:2. The annual amount of excrement for each animal is 9000 kg. The Nutritional elements in cattle manure is

	Organic matter %	*(N%)*	*P_2O_5(%)*	*K_2O(%)*
Faeces	14.0	0.3	0.2	0.1
Urine	2.3	1.0	0.1	1.4

Caudal fin The tail is also known as the caudal fin. Its fleshy muscular part where the tail joins the body is the caudal peduncle. The tail fin is the most important of all the median fins, being used by the fish for moving forward when swimming. The caudal fin is highly developed in most fishes and is variously modified. There are three major types of caudal fins, namely, protoceral, heteroceral and homoceral.

Caudal peduncle The hind region of the body from the anal fin to the base of the tail fin.

Cauliflower Disease Massive papillomatous lesions usually on the snout. Papillomas vary in shape and size from low elevations to extensive nodular or leaf like folds of tissue projecting above the skin surface. They may be single or multiple soft or relatively firm in consistency but do not metastasize. Colours ranges from pinkish or red if there is a marked vascular supply, to grey brown or black if melanocytes are numerous. Several possible causes have been reported e.g. external parasites, virus like particles and a polluted environment, but their relation to the aetiology of the tumours is unknown. It occur or many species of fish from both fresh water and marine habitats in widely scattered geographical areas.

Centre Line Longitudinal axis of a canal dike, on a plan a line drawn along the centre of a particular structure dividing into two equal sized parts.

Centre/Central Wire Axial number of the strand about which wires are laid.

Cephalothorax In crustacean and arachnids, the anterior part of the body, formed by the fusion of the head and thorax. It is connected to the abdomen and bears the mouth parts and walking legs.

Ceratophyllum/Horn Whort It is a submerged aquatic plant but not rooted to the bottom. Stem long branched soft with short internodes, roots absent, leaves 3 forked, filamentous, rigid, grow in whorl, flowers small sessile.

Cercariae Free swimming larvae of digeneans, these stages are usually released from the gastropod intermediate host.

Ceriodaphnia Belongs to phylum Arothropoda, order cladocera family Daphnidae. Head without a beak small and depressed 1st antennae short valves oval or rounded ending in a sharp short dorsal spine.

Cestodes/Tape Worm *Lingula intestinalis* Fishes infected shows sluggishness with distended abdomen and appear dull, sickly and with parts of the alimentary canal swollen or completely choked with cysts.

CFTRI Central Food Technological Research Institute is responsible for research in different types of food processing technology, and matters connected with sea food processing and packing. It is recognised by FAO as international centre for training in Food technology. The head office is at Mysore.

Chalimus The modified copepoded stage of a copepod parasite, attaches to its host by a filament.

Chanda An older name for the genus Ambasis.

Channa Genus belonging to the family of snake head fishes or ophiocephalidae.

Chara/Stone Wort It is a floating aquatic plant, stem or thallus of the plant soft, erect and branched, roots multicellular rhizoids, leaves simple or branched thread like in whorl with black spots, flowers absent.

Charax The pointed stake referring to the teeth.

Chela oxygastroides C.n. The glass barb. i.c. The body is elongated laterally compressed body structure. The dorsal contour is nearly straight while the belly edge arches outwards to the insertion of the anal fin with a sharp bow, D-9, A 31-35, no barbels length upto 8 inches, the colour is glassily translucent. Dis. In all kinds of standing and flowing water in Indo-China and Indonesia.

Chemical Oxygen demand (COD) The oxygen equivalent of the dichromate consumed during oxidation of a raw water sample. The chemical oxygen demand is a measure of organic matter and often permits an estimation of the BOD.

Chemotherapy Treatment with chemical or drug to remove parasites or other disease carrying organisms.

Chemotrophism A type of nutrition in which the source of energy for synthesis of organic requirement is chemical. Most chemotrophic organisms are heterotrophic (i.e. chemoheterotrophic) and their energy source is always an organic compounds, animals fungi, and most bacteria are chemoheterotrophs. If autotrophic, the energy is obtained by oxidation of an inorganic compound.

Chi Square test A stastical test used to determine if a set of observed frequencies differs (to a degree that would be improbable by chance alone) from those expected on the basis of a specific hypothesis.

Chikuwa (Japan) Fish jelly product of a cylindrical shape made by wrapping kneaded fish meat around a stick then baking it in a very hot oven. Similar product (made from cheaper fish) Yaki-Chikuwa.

Chill Storage Storage at temperature of melting ice (0°C).

Chilled Fish Fish kept at or close to the temperature of melting ice (0°C/32°F) but not frozen. Also called wet fish.

Chilodonella A colourless, oval flattened organisms which creep rapidly over the fins, body and gills. Under high magnification faint bands of cillia can be observed. This is highly destructive.

Chironomid An insect of the order Diptera, Whose young ones are blood red and worm like invariably found at the bottom of ponds living in tubes.

Chitin A nitrogen containing heterosaccharide found in some animals and cell walls of most fungi. The outer covering of arthropods the cuticle is impregnated in its outer layers with chitin which makes the exoskeleton more rigid. It is associated with Protein to give an uniquely tough yet flexible and light skeleton; which also has the advantage of being water proof. The chitinous plates are thinner for stiffness as required. The plates cannot grow once laid down and are digested at each moult. Chitin is also found in the hard parts of several other groups of animals. Chitin is a polymer of N-acetyl glucosamine. It consists of many glucose units, in each of which one of the hydroxyl group has been replaced by an acetyl amine group (CH_3CONH).

Chlamydomonas It is a Phytoplankton belongs to family chlamydomonadinae. It is microscopic, unicellular spherical or pyriform in shape with anterior end pointed and bearing two flagella, chloroplast horse shoe shaped.

Chlorampenicol/Chloromycetin It is a broad spectrum antibiotic effective against most gram negative and gram positive bacteria, particularly Aeromonas, Pseudomonas and Vibrio.

Chloride cells These cells contain granules, which take acidic stain hence called acidophilic. They are provided with a large number of mitochondria. These cells are also known as ionocytes because they help in the ionic regulation in he euryhaline and stenohaline fishes.

Chlorination The application of chlorine to water, sewage or industrial wastes for disinfection or other biological or chemical results.

Chlorine Greenish yellow pungent smelling gas which acts corrosively on mucous tissues. It combines with hydrogen into colourless hydrochloric acid. Cold water retains more free chlorine gas than warm water but it is nearly completely eliminated if the water is left standing for a few days.

Chlorinity It is defined as the total amount of chlorine, bromine and iodine in grams contained in one kg of Sea water, assuming that the bromine and the iodine had been replaced by chlorine. Ordinarily the chlorides are determined by titration with silver nitrate. Since the quantitative relations of major elements in sea water are remarkably constant the salt content the so-called salinity of sea water can be computed from chlorinity.

Choke It is a triangular piece at wing end. It is necessary in purse seine to get the Float line and load line while having the net by power block.

Chromatophores One of the pigment cells usually found on the skin of fish.

Chromophores Dermal cells which contributes to skin colour

CIBA Central Institute of Brackishwater Aquaculture conducts research and training in brackishwater acuaculture including shrimp farming, hatchery management and production. The Headquarter is at Madras.

CICEF Central Institute of Coastal Engineering for Fishery conduct engineering studied related to the construction of Fishing harbours and other coastal structures for fishing. It provides expertise in aquaculture engineering. The Headquarter is at Bangalore.

CICFRI Central Inland Capture Fisheries Research Institute conducts research on capture fisheries with respect to inland water area (river, lakes, reservoirs, estuaries and backwaters). The Headquarter is at Barrackpore, West Bengal.

CIFA Central Institute of Freshwater Aquaculture under takes research and training in culture and farming of freshwater carps, cat fishes, prawns, snails, and other aquatic organisms of commercial importance. The Headquarter is at Bhubaneshwar.

CIFE Central Institute of Fisheries Education is a deemed university imparts graduate and post-graduate level training on all aspects of Fisheries. The Headquarter is at Bombay.

CIFNET Central Institute of Fisheries Nautical Engineering and Training. It is responsible for training of personel for managing fishing vessel and operating crew including engine drivers and deck hands. The Headquarter is at Kochi.

CIFT Central Institute of Fisheries Technology undertakes research on all aspects of fishing and post harvest technology. The Headquarter is at Cochin.

Cilia Microscopic hairs in and on animals, used in feeding and movement.

Cilium A locomotory organelle found throughout animal kingdom composed of a mircotublar axoneme and covered by the outer cell membrane.

Cirrus Coarse cilia which are longer than normal.

City Farm A theoretical concept in which all the wastes from an urban community are processed into forms that can be fed to fish or to livestock.

CL Carapace length, usually measured from the eye notch to the posterior mid dorsal margin of the carapace.

Cladocera Fish food crustaceans belonging to class phyllopoda. The body is generally enclosed within two shells and little articulated. Two lateral eyes which on the final, completely developed form fuse into one unpaired eyes *e.g.* Daphnia, Bosminae.

Clam Liquor Liquid extracted during the cooking and opening of clams, also called clam extract, undiluted it is called clam juice, diluted is called clam broth, and when concentrated by evaporation is called clam nectar may be canned in all these forms or used to fillup canned clam meat.

Clarias batrachus (L) Body structure like as description in family. D/62-76, A 45-58 length upto 20 inches. Gray blue with lighter spots fins green with red edges.

Clariidae Family of Gill sac catfishes. The body elongated shape with flattened head and terminal mouth around which they carry four pairs of barbels. The front part of the body is little compressed laterally but the rear part is strongly so and nearly ribbon shaped. The body is naked and covered with a leathery mucous skin. All fins are without spines. An adipose fin may be present. The gill sac catfishes are inhabitants of biologically very unfavourable waters, which they are able to leave for hours and even days.

Claspers The copulatory organs of male cartilaginous fish. They consists of a pair of grooved rod like projections; one on each side of the cloaca between the Pelvic fins.

Clay Extremely fine textured soil made up of small mineral particles less than 0.002 mm in diameter, clay soils become very sticky and difficult to work when wet and can easily become waterlogged. Nutrient availability to plants can be a problem as the nutrients may become chemically bound to the surface of the particles.

Clay Loam This is a fine textured soil. The characteristics of which are midway between the characteristics of the silt loam and the clay loam. Its moisture conditions are ideal. It is possible to form a ribbon with it when squeezed between the thumb and finger.

Cleansed Shellfish Live molluscs such as mussels from which polluting bacteria have been removed by immersion in sterile sea water.

Clinometer It is a simple rugged instrument for determination of the required rope length for towed devices which shall be operated at a certain depth. The angle of inclination of the towing rope is determined by taking bearings through the slit of the clinometer.

Clitellum The portion of a leech which is located between the trachelosome and urosome, it secretes the cocoon.

Cloaea The single chamber into which the alimentary canal, kidney ducts and genital ducts discharge their contents.

Clone A line of cell derived from a single cell. All cells of a clone remain genetically identical.

Closed Mating System A breeding programme in which no outside individuals are allowed to introduce gametes into the population. Hence the progeny are produced by known parental combination.

Closure Regutations that prohibits fishing activity in specified fisheries. It is specified to a particular time period and or a particular place.

CMFRI Central Marine Fisheries Research Institute, undertakes research on all aspects to Marine fisheries including studies on resource potential and fish species, population dynamics and mariculture. The Headquarter is at Cochin.

Coagulation Necrosis It is the result of ischaemia or loss of blood supply to an area. It can be recognizable in sections as an acidophic area where the cell nuclei are destroyed but the outline of the cells are still visible. It is most common in the centre of rapidly growing tumours or tubercles.

Coaltar A preservative for heavy nets, ropes etc. It gives more physical protection for the nets against wear and tear.

Coastal plain A plain that borders the sea and extends inland to the nearest elevated land.

Cobble Stone Rounded or partially rounded rock or mineral fragments between 3 and 10 inches in diameter.

Cocoon A brooding capsule for (leech) eggs.

Cod end The posterior end of a collecting bag.

Codend knot Knot tied to fasten the cod and rope after closing the tail of the cod end.

Cod Liver Meal Made on a very small scale in some areas by drying the residues from cod liver oil manufacture. Used as animal feeding stuff, in Japan also made from other species than cod.

Cod Liver Oil Oil extracted from livers of cod and sometimes other suitable gadoids such as heddock.

Cod Liver Paste Edible paste made from cod livers with spices and other flavouring ingredients.

Coefficient of Twist It is a measure of the twist hardness of yarn, determined by multiplying the turns per unit length by the square root of the count in a direct system.

Coelozoic Parasites Parasites which occur in organ cavities.

Coil A continuous length of rope or line arranged in the form of a spiral.

Coir Strong thick fibre obtained from the husk of coconut fruit with which ropes, cords etc. are made.

Cold Smoked Fish Fish cured by smoking at air temperature not higher than 30°C (86°F) in some countries 90°F to avoid cooking the flesh or coagulating the proteins, tropical species may be smoked at slightly higher temperatures.

Cold Storage Storage at temperatures below freezing point, generally below-18°C at (0°F) in many countries recommended temperature is below-30°C.

Cold Water Disease/Penduncle Disease Its affect young fish of most of the cultured salmonid species with occasionally occurring acute mortality . Infection usually develops at temperatures between 4° and 12° with losses falling of sharply if the temperature rises. Often the worst affected groups are those most densely stocked or with most organic debris present in the holding facility. The gross lesions are normally confined to the fins, skin and muscle.

Coliform Bacteria A group of bacteria which are normally abundant in the intestinal tracts of man and other warm blooded animals and are used as indicators (no. of individuals per ml. of water) when testing the sanitary quality of water.

Colisa Genus of Gouramis, belonging to the family of climbing perches Anabantidae. The body is round elongated and strongly compressed laterally. The dorsal fin is as long as the anal, the second ray of the ventrals has grown into a long thread like formation.

Colisa fasciata (Bloch) The giant Gourami. Besides the description in genus, the mouth small possesses thick fleshy lips D XV-XVII/9-14 A XV-XVIII/14-19, the scales are relatively large. The colour is reddish to brownish base with blue green cross strips. dis-India to Burma.

***Colisa labiosa* (Day)** The thick lipped Gourami. The small terminal mouth is located beneath a remarkable thickening with results from a puffing up of the skin above the upper jaw in males, D XV-XVIII/8-10 A-XVI-XVIII/17-20 length upto 4 inches. dis-Indo-China.

***Colisa lalia* (Ham)** The dwarf Gourami. It is smallest and shortest species. D-XV-XVIII/7-10 A XVII-XVIII/13-17 length upto 2 inches. The colour is shiny blue green base with slanting vivid red cross bars the fins shiny blue and green with red dot and seams. dis—India.

Collagen The principal proteinaceous substance in connective fibres.

Colloid A minute division of the soil either mineral or organic (made up of finely divided humus) without a colloidal fraction soils are physically dry and chemically inert. The vast surface area of colloids has electrical attraction for mineral ions released during weathering or contained in soil water. As a result colloids hold a vital reservoir of exchangeable plants nutrients.

Colour malformations Divergences from the natural colouring of fishes which even if not directly fatal must at least be considered malformations.

Column feeder A fish that generally feeds at the surface or in mid deep water.

Commensal Organism which lives on, in or within another may offer some advantages but does not cause harm.

Commercial Banks Financial intermediaries that accepts deposits from households, firms, governments, foreigners and other financial institutions. They also make loans and buy negotiable debts.

Commercial harvesters Harvesters who receives a significant proportions of the catch / income.

Commission That portion of the premium paid by the insurer to the agent as remuneration for services in procuring and serving the policy holder.

Common fish diseases

Sl. No.	*Diseases*	*Factors associated with diseases*
1.	Furunculosis	Crowding, bad handling and low level of dissolved oxygen in water
2.	Bacterial gill disease	Crowing, low level of dissol- -ved oxygen and higher concentration of ammonia in water
3.	Corynbacterial kidney disease	Diets containing corn-gluten and low total hardness (less than 100 ppm as $CaCo_3$) of water
4.	Columnaris	Crowdding, bad handling and higher water temperature
5.	Aeromonas & Pseudomonas hemorrhagic septicemia	Low dissolved oxygen in water, handling, crowding, prior sufferings from infection of *Costia* or *Trichodina;* non-sanitary condition of ambient water together with more bacterial load; Chronic exposure to pesticides.
6.	Coldwater	Low temperature of ambient water
7.	Spring	viremia of Bad handling associated with low temperature
8.	Nephrocalcinosia	Overstocking
9.	Gill necrosis	Over drugging with formalin
10.	Cataract of *Catla* catla	Poor sanitary condition of ambient water
11.	Dropsy of carps and catfishes	Poor sanitary condition of ambient water, over stocking and chronic exposure to low level dissolved oxygen.
12.	Tumours of *Anabas testudineus*	Same as above
13.	Reddish blotches of *Hypophthalmichthys molitraix*	Same as above

14. Intectious hemato poetic necrosis	Low temperature of ambient water
15. Vibriosis	Bad handling, low level of dissolved oxygen in ambient water .
16. Tail and fin rot	Crowding, imporper diet, temperature and chronic exposure to PCBS
17. Blue-sac	Crowding; accumulation of nitrogenous metabolites in ambient water
18. Gas-embolism	Supersaturation of O_2, N_2 and Co_2 in water
19. Trichodinosis	Crowing, low level of dissolved oxygen in water.
20. Coagulated yolk	Rough handling, mineral loss in ambient water
21. Handling loss	Repeated netting-cum-rough handling.
22. Fatty infiltration of liver	Deficiency of essential fatty acids in diet
23. Improper growth	Deficiency in essential amino acids in diet, poor sanitary condition, etc.
24. Expothalmia	Deficiency of vitamin E in the diet
25. Anaemia and cloudy lens	Deficiency of riboflavin in the diet
26. Scoliosis and lordosis	Deficiency of ascorbic acid in the diet

Common Property When the usage of property become unrestricted, everyone has right to utilise the resources.

Compaction Increasing the density and lowering the porosity of a soil by mechanical manipulation essential in the construction of ponds.

Compatible species Fishes which can live and grow together in a pond without preying upon one another or competing among themselves for their food.

Competition Interactions between two or more organisms or types of organisms attempting to use the same resources that is limited in availability.

Competition Index A measure of the change in yield due to adopting polyculture of different species instead of mono-culture with single species.

Complete Diet A diet capable of supporting good growth and survival throughout a specified culture phase, e.g. larval life or on growing.

Composite Fish Culture In order to obtain high production per ha. of water body, fast growing compatible species of different feeding habits, or different weight classes of the same species are stocked together in the same pond so that all its ecological niches are occupied by fish.

Composition of Fish muscle protein 15-24% lipid 0.1-22%, Carbohydrate 1-3%, Inorganic matter 0.82% and water 66-84%.

Table : Vitamins and mineral composition of some fish

	Vitamins mg/100gm				Minerals mg/100gm			
	A	D	B_2	Niacin	Ca	I	N_a	P
Horse Mackrel	20	0	0.16	4.4	65	0.5	120	250
Eel	4700	150	0.45	3.7	95	1.0	65	230
Skipjack	17	420	0.16	19.0	10	1.9	44	270
Salmon	200	210	0.17	8.4	14	0.9	95	210
Mackerel	100	330	0.54	9.7	22	1.5	80	160
Herring	300	40	0.29	4.0	100	1.1	100	260
Sardine	60	530	0.36	7.7	70	1.7	360	200
Oyster	55	–	0.32	2.0	55	3.6	280	130

Compost Decomposed organic materials. The nutrients of the compost are quickly absorbed by the phytoplankton. They consume less dissolved oxygen.

Conchiolin A nitrogenous albuminoid substance which is usually dark brown in colour, it is the organic base of many molluscan shells.

Condensed Fish Solubles The aqueous portion of the press liquor produced during manufacture of fish meal, from which some of the moisture has been evaporated to give a thick syrup containing usually 40-50% of solids, may be marketed as such or added back to press cake before drying to give a whole fish meal or full fish meal.

Condition Factor Weight/length relationship.

Conditioning Process of keeping young fish in a limited space for sometime to get the gut, voided of contents and also to a accustom them to the limited space.

Conductivity The conductivity of water is a measure of its ability to convey an electrical current. Different ions vary in their ability to conduct electricity but greater the concentration of ions in water the larger the conductivity.

The unit of conductivity is μ mhos/cm. Distilled water conductivity is 1 μ mhos/cm while natural water have conductivity of 20-1500 μ mhos/cm. Conductivity meters are manufactured by several companies.

Conjugation A sexual process in ciliates in which two individuals fuse their micronuclei, undergo meiotic division followed by exchange and fusion of the haploid nuclei in the conjugants

Conservation Careful management and protection of natural resources and the environment for sustainable benefit from them.

Consumer Related Risks Risks due to loss of consumer appeal health regulations action of third party (the consumer).

Consumer's Surplus The difference between the value which the consumer places on a goods and the price which the consumer must pay for the goods.

Continental shelf The submerged extension of continental masses extending seaward from the shoreline to the 200 m depth contour.

Continental Climate Any climate in which the difference between summer and winter temperatures is greater than the average range for that latitude because of the influence of a large land mass.

Continuous Culture A system of rearing prawn to ponds which involves continuous pond operation, ponds are not drained for harvesting nor completely harvested. The larger animals are regularly removed by seine net for marketing, leaving the smaller ones to grow. The ponds are regularly restocked with post larvae or juveniles.

Continuous Filament Fine man made fibres indefinite, practically infinite length.

Contour A line drawn on a map joining all places with the same height above or below mean sea level. The relief of an area can be shown on a map by drawing a series of contours representing different heights.

Contour Mapping Contour line show the configuration or changes in elevation of the ground in a topographic map.

Each contour line represents points of same elevation and are spaced according to the difference in elevation between two adjacent lines, e.g. the contour line (C.L.) 0.50 means that every point on that line has elevation of 0.50m above the reference datum. If the next contour lines lower and above C.L. 0.50 are 0.25 and 0.75 respectively, the contour interval is 0.25.

Contour Ponds It is constructed along the side of the valley, the bottom of the pond lies along the contours of the land.

Cook Chill Process Process in which the food is fully cooked, immediately chilled, stored at low temperature, above freezing point (0-3°c) and reheated appropriately before serving.

Cooperative Society A cooperative society is defined as an association of persons varying in number who are grappling with the same economic difficulties and who by joining together on a basis of equal rights and obligations endeavour to solve these difficulties mainly by conducting at their own risk. A joint undertaking to which they have transferred certain economic functions corresponding to their common needs and by utilising this undertaking jointly for their common material and moral benefit.

Copepodid The larval stage of copepod, it is between the nauplius and the pre-adult stages.

Copepods Small planktonic or bottom dwelling crustaceans that are important components of the food chain of aquatic animals.

Copulation Introduction of the spermatozoans into the female body in order to achieve fecundation with the aid of special organs. In fishes real copulation only take place in special cases.

Copulatory organs The copulatory or intromittant organs are used for introducing sperm into the reproductive tract of the female. They are generally absent in fishes as fertilization is predominantly external. However, in certain fishes copulatory organs like claspers and gonopodium are present.

Coracle It is extensively used in Tanjore and along the course of the Cauveri. It consists of a very large wide mouthed circular flat bottomed basket. The sides are comparatively low.

Coral Fish Disease Epidemic disease of coral fish and other salt water fishes from tropical regions, the cause of which belongs to the same genus Oodinium, the unicellular algae as the causer of the well known velvet disease. The germs generally attack the gills. High temperature help to spread the disease further. The cycle of development resembles to ichthyophythiriasis.

Coral Reef Stony formation build up from the seafloor by colonial coral polyps.

Cordage All kinds of ropes lines cords and twines manufactured from natural or man made fibre.

Core Filling element in the centre and along the axis of a strand (strand core) or a rope (main core) serving mainly as support for the structure.

Core Mantle Rope This rope differs from the normal tubular and solid braids in that the load carrying capacity is more in the core and less in the braided cover, which serves only to keep the core together and ensure for stability of the rope.

Corixa/Water Boatmen Insects belong to order Hemiptera, family-corixidae. Body is dark greyish often mottled faint cross lined with yellow middle legs extra ordinarily long and hind legs are flattened and fringed.

Cormorant A species of water bird eating fish which dives into catch fish.

Cornified Containing fleshy fibrous tissue.

Cosmoid scales Such scales are not found in any of the fishes living today but they were the type characteristic of extinct lung-fishes. Their body was covered with thick cosmoid scales of rhomboid shape and the scales did not overlap. Each scale consisted of vitrodentine, cosmine, vascular layer and isopedin layer. In the modern Dipnoi, the cosmoid scales have been greatly reduced, being thinned to a single layer of bone.

Cost Benefit Analysis A method of allocating scarce resources efficiently by equating marginal social costs and benefits.

Cost of Input Per Unit of Production It can be expressed in cost per kg, Man hour per kg, feed per kg etc.

Costiasis This disease is produced by *Costia necatrix* is the most common mastigophoran infection observed in ponds fish. It is pear shaped organisms can be detected from their movement which resembles a flickering flame or by a careful examination of the margin of a fin.

Cotton Seed fibre obtained from cotton plant. Gossypium is more extensively used for netting purpose than any other plant fibre.

Cotton Wool Diseases It is common in tropical aquarium fish.

Count This term used by shrimp buyers refers to the number of prawn or prawn tails per unit weight when using this form it is important to state whether shell-on/head-on, shell-on tails or peeled tails are being described.

Creasing Increasing the number of meshes in successive rounds to alter the shape of a webbing.

Creek A small stream of water that serves as the natural drainage course for a drainage basin of nominal or small size. Its size is relative small, some creeks in the humid section would be called rivers if they occured in the arid portion.

Creel limits Regulations that restricts the number of fishes a single participant can harvest daily in a recreational fishery.

Crest The top of a dam, dike, spillway or weir to which water must rise before passing over the structure.

Critical Standing Crop (CSC) When natural food is fully utilised for maintenence and maximum growth of fish.

Crustaceans Arthropods with bilateral symmetry, exoskeleton, and two main body segments (cephalothorax and abdomen).

Cryo-preservation Specialized process for freezing microscopic organisms or gametes for long term storage in a dormant condition. Normal life function and resumed on subsequent thawing.

Ctenoid Scales Scales with a more or less deeply toothed edge also called comb scales.

Ctenopharyngodon Idella c.n. Grass carp, white amur, dis. Naturally found in Amur region. North China to South China, USSR. lower reaches of river Amur, Introduced in India. i.c. Body moderately elongate, sub-cylindrical anteriorly and compressed posteriorly, Abdomen rounded. Head depressed and flattened mouth terminal Lips thin without barbels.

Cubic Feet Per Second (CFS) A unit expressing the rate of discharge of a liquid through a rectangular cross section. One foot wide and one foot deep flowing at an average velocity of one foot per second.

Cull Partial harvest of (usually) largest individuals in a population.

Cultivable Which can be cultivated and could be grown under controlled conditions.

Cultural eutrophication Nutrient enrichment that occurs due to activities of humans. Common source of nutrients enrichment include effluents from municipal sewage treatment facilities and fertilizer runoff.

Culture Process of cultivation of fish in all its stages.

Cuprinol A copper naphthenate solution used as a preservative for all kinds of twines and nets.

Curing The processing of fixing the tannin in the twine after the tanning of the net.

Current A Large mass of continuously moving water.

Cutch Tannin or extract taken from the bark and wood of certain tree used for preservation of nets against rotting some fruits and nuts are also used for tanning nets.

Cutting Netting to Shape Cutting from netting pieces in the shape of trapezia, triangles, parallelograms or other polygons. (i) Bar cut : Cut parallel to the line of sequential mesh bars .(ii) K-cut (Knot cuts) : Cut just beyond the knots where the two legs are cut at a time. (iii) N-cut : Cut at right angles to the general course of netting. It is known as point cut. (iv) T-cut : Cut parallel to the general course of netting. In general this is known as mesh cut (N-cut).

Cutting Rate The sequential cutting of meshes to reduce netting width at a given rate.

Cybisterlarvae/Water Tigers/Dragons of the Pond Muddy green in colour, body long, slender and segmented, swim by up and down movements of body can be seen hanging from the water surface film.

Cybrister/Diving beetle It is an aquatic insect belongs to order coleoptera. It has shining black or brownish black marked with dull yellow margins, Broad, tightly fitting and capable on only slight movement, antennae small and slender and hind legs long with hairs, wings functional. It grows upto 35 mm.

Cyclops Belongs to phylum arthropoda order cyclopoida, family cyclopeidae. Body elongated, head and thorax united, antennae shorter than cephalothorax 4 or 5, 2-branched, swimming feet on the thorax, abdomen without appendage, female carries 2 egg sacs.

Cyclostomatal (Gr. Cyklos = Circular, Stoma = mouth). It includes the modern agnathans which are parasites and scavengers in the adult stage. Their endoskeleton is cartilaginous and they lack an exoskeleton. Their Cylindrical bodies are smooth and slimy. The anterior end is modified either into a sucking funnel or into a depression edged with tentacles. The tongue bears horny teeth. Larval stage is microphagous.

Cyonophyceae Blue algae Algae belonging to the class of the Schizophyte. Quickly proliferating algae, very obnoxious which form blue green coloured carpets with an evil smelling veil.

Cyperus Superficially resembling the grasses and differing from them as the stem is triangular and solid with leaves in 3 ranks and with close sheaths, spikelets flattened the seales in 2 rows on opposite sides of the rachilla.

Cyst A non-motile resistant dormant stage.

Cystacanth A juvenile acanthocephan that develops from an acanthella, it possesses a proboscis and is ineffective to the definite host.

Cystogenous glands Glands which produces encystment substances.

D

D-Shackle Shackle with straight parallel sides.

Dactylogyrosis An infection of Dactylogyrus spp. causes the fish to secrete more mucus, in addition gills become pale the operculum opens, dyspnoea occurs and there is evident dropsy of gills. The infected fish swims slowly and is anemic.

Daphnia A fresh water crustacean (cladocera) also called a water flea.

Dead Spot Area or volume of pond bottom or water where circulation is minimal and where sedimentation and anaerobic condition develop.

Debris Rubbish garbage anything that is not supposed to be in certain area (pond).

DecapodsThe best known order of the crustacean comprising the crustacean most often consumed by lobster, crab, shrimp. The decapods have the fused cephalothorax covered by a chitinous carapace and the five pairs of legs. The arrangement of the appendages into swimming and walking legs varies with the individual species.

Decapsulation Removal of the outer shell of Artemia cysts by dissolution in chlorine solution.

Dechloration Removal of chlorine from chlorinated water. It can be done by continuous aeration or application of chemicals (*e.g.* sodium thiosulphate).

Decomposers Some multicellular animals fungi and bacteria that utilize the tissues of dead animals and plants for their own metabolism needs. They eventually break down complex proteins into simple materials that are essential for the plant growth.

Deep Sea A sea zone at depth below approximately 3000m on the 4°C isotherm, the sea zone at depth below 6000m is hadal. The depth of these deep sea zones is set by international agreement. There is usually no plant life at depth where there is no light. The animals in the deep sea are adapted to both pressure and darkness. They represent all major invertebrate phyla and many fishes. The deeper an organisms environment the greater the likelihood that it is a detritus feeder living on pieces or dead plants and animals or a deposit feeder.

Defence Mechanisms The means by which organisms defend themselves against predators. Some organisms use camouflage or move rapidly. Some clam up and others particularly slow or nonmotile organisms, uses chemical defences. For example some brown algae use chemical defences to make them unpalatable to predators. Several species of sole produce shark repellants and some sessile (stationary) organisms produce toxic substances.

Defoliant Herbicide which is designed to remove leaves from trees and shrubs or to kill plants. Such defoliants as 2, 4-d and 2,4, S-T disturb hormonal balances in plants and so induce metabolic disorder.

Degree Day The departure per degree per day in the mean daily temperature from an adopted standard references temperature usually 65° F. The number of degree days for an individual day is the departure of the mean temperature from the standard and the number of degree days in a month or other interval is the sum of all the daily values.

Dehydrated Fish Originally fish that had been dried under controlled conditions to a predetermined moisture content as opposed to fish that had been dried by exposure to natural climatic conditions or by use of mechanical drying equipment.

Delayed hypersensitivity A cell mediated immune response producing a cellular infiltrate and odema that are maximal about 48 hours after antigen challenge.

Delta Fan-shaped feature formed by sediments that pile up at a river mouth.

Demand/Pendulum Feeder Type of feeder in which the fish are able to nudge a pendulum which operates a gate and releases food in the water.

Demersal Living near the bottom.

Demersal Trawl See bottom trawl.

Denatant Swimming, drifting or migrating with the current (usually eggs and larvae).

Denatured protein A protein that has lost its natural configuration by exposure to a destabilizing agent such as heat.

Dendritic Drainage The form of the drainage pattern of a stream and its tributaries when it follows a tree like shape with the main trunk branches and twigs corresponding to the main stream tributaries and subtributaries respectively of the stream.

Denitrification The chemical reduction of nitrate to nitrogen by certain microorganisms.

Density dependent Characteristics of fish populations that are subject to the influence of population density or abundance e.g. the growth of fish is inversely proportional to population abundance.

Dependent Variables A variable which is influenced by changes in another variable. In an equation relating expenditures is an income, expenditure is the dependent variable being influenced by changes in income.

Deposition The settling of suspended materials from the water column of flowing water. The settling usually occurs as the rate of flow of water declines.

Dermatitis The inflammation of the skin.

Derries powder A commercial insecticide made of the powdered roots of the plant *Derries elliptica* used as a fish poison.

Desalination The process of removing salt and other dissolved minerals from sea water to make it useful for agriculture industry and human use.

Detritus Freshly dead or partially decomposed organic matter.

Diadromous The life history of fish with both freshwater and marine phases.

Diaptomus Belong to Arthropoda group, body long and pear shaped but without shell like covering, 4, 5, branched feet. 1st antennae as long as body having 23-25 segments, egg sac single.

Diatoms Single-celled protists (algae); part of phytoplankton community.

Diet A specific type or combination of prescribed food.

Diffusion The transfer of matter or energy from a region of high concentration to one of lower concentration. This is the mode of transmission of light through the photic zone of the ocean.

Difinitive host The host in which the parasite usually undergoes sexual reproduction.

Digestibility The digestibility represents the percentage of nutrients absorbed by the fish.

$$\text{Digestibility (\%)} = \frac{\text{Nutrient intake - nutrient remaining in faeces}}{\text{Nutrient intake}} \times 100$$

Digestibility is affected by many factors and it is possible for the same ratios to have various digestibility. Enzymes activity is associated with temperature and digestibility rate will increases, when the temperature rises with in the adaptable temperature ranges. The presence of crude fibre in the diet will reduce the digestive rate of other components. The amount of food consumed by the fish will also affect the food digestibility, a proper food intake will maximize the digestibility.

Digestion Rate It is the time taken to digest the food i.e. the rate of passage of the food through the digestion tract. This has been studied mainly in fish with a stomach, where the amount of food in the stomach was determined at various times after feeding. Digestion rate seems to be independent of the size of the fish. However other factors may affect it. Food composition is one of these factors. Different digestion rates have been recorded in fish feed.

It is reported that the algae, microcystis, Anabaena, chlorella were retained in intestine for a longer time than spirogyra and ulothrix. The digestion rate of fish with stomach is usually slower than in fish without stomach.

Dike A high barrier built along the banks of river system / Ponds to prevent flooding or flood plains during high water periods of the year.

Diminishing Marginal Returns (law of) When additional units of one inputs are used with other inputs held constant, a point is reached beyond which marginal product begins to decline.

Dimorphic Existing in two morphologically different form

Dinospore A unicellular haploid motile stage of dinoflagellate parasites that resembles a free living dinoflagellate.

Dip A treatment or disinfection bath into which one or more animals are placed for a short time (1-6 minutes).

Diphycercal tail The type of tail found in lampreys, hay fish. The young stages of all fish and the larvae of amphibians. The vertebral column extends into the tail, which has caudal fins of equal sizes above and below it.

Diploid Having two sets of chromosomes i.e. the number carried by all body cells.

2.4.-D; 2, 4,-D; 2, 4, Dichlorophenoxy acetic acid, a synthetic plant growth regulator with auxin like activity used as a herbicide.

Diplostomiasis/Black Spot Disease It is caused by Diplostomulum sp. Small black nodules of about 1-5 mm diameter in the affected region is the common symptom.

Diplozoic Having two karyomastigonts.

Dipnoi Lung fishes, found in the freshwater they are characterized by a functional lung and internal gill as well as external nostrils allowing air breathing at the water surface without opening the mouth. The heart and blood system are adapted for pulmonary respiration and resemble those of amphibians.

Diquat Contact herbicide used to control broad leaved weeds in many situations including still and slow flowing water, it is harmful to mammals.

Directed catch The portion of any catch that is composed of the texa specially being pursued during the fishing activity.

Discharge Capacity Rate of flow or the volume of water pumped per minute cubic feet per second.

Discharge Head It is the vertical distance from the central line of the impeller to point of discharge.

Disinfection Reduction without adversely affecting the fish by of hygienically satisfactory chemical agents and or physical methods of the number of micro organisms to a level that will not lead to a harmful contamination of fish.

Disporic A small sporogonic plasmodium or pseudo-Plasmodium which produce only one spore.

Dissolved Oxygen The oxygen that is dissolved in water.

Distilled Water Water that has been made chemically completely pure by evaporation.

Diurnal Oscillation The term referring to variations in the planktonic population in the course of a day. Plankton moves up toward surface in day time and down at night.

Diversion pond This type of pond is supplied with water through a supplying canal which diverts the water from a river or some other type of water body.

Doctor fish The tench *Tinca tinca* so called as its body slime is believed to have a curative effect on wounds.

Dog Ear Triangular pieces of netting fixed into the angle formed by the forward edge of a trawl having all bars along the hanging end points on the wings.

Dol Net It is similar to a trawl net in looks, it differs from the latter in that it is a stationary gear which makes use of the tidal energy to catch fish. It is most commonly operated along the coasts of Gujarat and Maharashtra in comparatively shallow waters.

Dorsal fin The unpaired fin along the middle line of the back.

Double Cropping Population of two crops from the same pond, simultaneously (polyculture) or alternately (crop rotation) one of the crops need not be an aquaculture product.

Double Mesh Mesh with all the parts made of double twins. This is given to those part of gear where special strengthening is necessary. Double meshes are made either by double needle or by single needle in the latter case by the repetition of making half meshes in successive rounds.

Double Sheet Bend Knot It is a similar knot of single sheet bend but with the end rope led twice underneath the eye instead of once.

Dragon Fly Nymph Muddy brown in colour, head broad and the lower side is concealed by the developments of the lower lip in form of a long jointed arm like structure, antenna short.

Draw Down It is the lowering of water surface below the static level during pumping.

Dredge A floating device used for removing underwater material and usually held in place by supports that descend to the bottom.

Dressed Crab White and brown meat extracted from the cooked whole crab, seasoned and laid out attractively in the cleaned carapace shell, marketed fresh or frozen, also canned (in Norway for canned dressed crab minimum crab content is 90% of weight).

Dressed Fish Fish ready prepared for cooking or special preparation for good presentation, also called pan-ready, kitchen ready fish.

Dried Fish Fish preserved by removal of sufficient moisture to retard or prevent the growth of bacteria and moulds. Bacterial activity ceases when the water content is less than about 25%. Moulds can not grow when the water content is 15%.

Dried salted Fish Fish preserved by the combination of salting and drying, applies, mostly to non fatty fish particularly cod, ling coal fish, haddock, hake and tusk.

Drug Resistance The ability of a microorganism to grow in the presence of a drug, usually an antibiotic.

Dry Bundh It is a shallow seasonal pond or just a depression in the land, bounded by embankments on three sides which impounds freshwater from the catchment area during monsoon season. It remains dry for the greater part of the year. The size varies from 0.2 to 2.5 ha. The catchment area varies from 5-25 times depending on average rainfall.

Dry Dock A ship yard structure used for major overhaul and repair to ships bottoms. The dry dock is a basin cut into a harbor and separated from the rest of it by a wall. A ship can be positioned over supports and the basin emptied, leaving the ship standing on the supports. After repairing and cleaning the water is allowed to flow back into the basin and the ship is refloated.

Duck weeds Group of floating plants belonging to the genera Lemna, Azolla, Wolffia.

Dy A type of lake bottom sediment composed largely of peaty plant detritus mixed with a gelatinous precipitate of iron salts. It is found in oligotrophic lakes.

Dyplasia The regressive change in adult cells

Dysnea Difficulty in breathing.

E

Earth Pond Elongated pond usually some times longer than wide made by simple excavation.

Ecchymisis A small haemorrhagic spot in the skin or mucous membrane.

Ecdysis The act of casting the external skeleton of crustaceans.

Echolocation The ability to determine the shape and size of an object and its distance by the echo, its make. The best known terrestrial animals who find their way about by echolocation are the bats. In sea the range of species using echolocation is much greater. All toothed-whales use echolocation. The sound used for echolocation is produced by several different structures in the animals head. In the whales it may (or may not) be amplified by the blow hole. The sound that is used, is specific to given species and a frequency range that is usually greater than human can detect. The sound receiver for echolocation is not always ear, even in those species the entire skull may be receiver and in some whales the lower jaw is an auditory receiver.

Ecosystem A term used to describe the interrelationship between all organisms in a given area and their relationship to the nonliving materials that make life possible.

Ectoderm The outer layer of cells of an embryonic organism. This layer in the embryo develops the nervous system and skin.

Eddy A fluid flow that is markedly different from the general movement of a large mass of that fluid. The number and velocity of the surrounding mass of fluid (air and water) increases.

Effective Rain Fall The portion of the total rainfall that reaches stream and rivers.

Egg membrane The thin outer envelope of egg which swells up after fertilization to form the vitelline membrane.

EIC/Export Inspection Council of India Eastablished to advise the Government in regard to measures to be taken up for quality control and pre-shipment inspection of various commodities exported from India. Under the EIC the export inspection agency (EIA) conducts pre-shipment quality control inspection and certification of notified commodities.

Eichornia It is commonly known as water hyacinth. It is a floating aquatic plant. The stem is modified rhizome roots adventitious fibrous leaves large with bladder like petiole flowers lavender colour in bunch.

Electrophoresis The seperation of proteins across an electrical gradient according to their mass and charge.

Elevers Upon reaching the coastal waters of Europe the Leptocephali metamorphosize into the elevers, also known as glass eels. The period of metamorphoesis extends to about a year during which the larva stop feeding, body becomes cylindrical in shape, the length of larva decreases the needle like teeth are lost and replaced by the newer ones. The elevers continue to develop in coastal waters for about three years. As they grow sufficiently stronger to swim up the rivers, they start swimming upstream to reach their parental home. Finally after reaching the home of their parents the eel like elevers become pigmented, which feed and grow for several years and become sexually mature before starting their breding migration to the sea.

El Nino Periodic warm ocean current that develops in middle of the Pacific.

Emaciation The deficiency of muscle and fat in an organism due to under nourishment.

Embroy The young one (fish) developing within the egg membrane.

Emprical Elementary, based on trial and error methods.

Emulsion A milky liquid prepared by mixing oil and water by means of soap.

Encapsulation The surrounding of one compound or substance with a thin coat of another. The term microencapsulation is used where the final product is very fine powder microencapsulation is used as a means of preservation.

Encystment The covering of a parasite with materials of parasite origin.

Endemic Fully restricted to a small geographic area or to single or a small number of water sheds.

Endocrinology The study of endocrine glands and their secretions (hormones).

Endocytosis The uptake by a cell of material from the environment by invagination of its plasma membrane.

Endospore The thicker inner chitinous layer of the wall in a spore.

English Knot/Sheet Bend Knot It is made by passing the end of one line through the eye or bight of the other around the eye and under itself.

Enteritis The diseased fish has an expanded abdomen with red blotches, the fins are congested and decayed, the annus is red and swollen and when slight pressure is applied to the abdomen a yellow mucus is released from the anus. The intestinal wall show hyperaenia and inflammation cells of mucus membrane ulcerate and drop off becoming bloody mucus and blocking the intestine. The diseased fish shows a loss of appetite, swims. slowly and alone and soon dies. The causative bacteria may be *Aeromonas punctata.*

Environmental Limits Those conditions that limit the range of an organism. They include temperature salinity etc.

Enzyme A protein catalyst produced within a living organism which regulates the rate of specific biochemical reactions.

Epidemiology Study of the distribution and determinants of infection or disease.

Epifauna Animals that live on the ocean bottom comprising both motile and sessile species.

Epigamic Character An animal character other than one associated with the reproductive organs which is concerned with sexual reproduction (e.g. bird song, distinctive coloration of many male fish.)

Epipelagic The term referring to the uppermost layer of the ocean water column down to a depth of about 200m.

Epilimnion Warm, well mixed surface water occuring above the thermocline during periods of stratification.

Epizootic An outbreak of a disease or an usually large increase in prevalence and/or intensity of a parasite in an animal population.

Epizootiology Study of disease in animal populations.

Equilibrium Biomass Long term mean biomass of a given species which can be sustained in a given region under defined and balanced conditions.

Eradication The complete extinction of a species throughout its range.

Erosion The removal of particles from substrates due to the abrasive action of flowing water / air.

Erythrodema It is a common disease in grass carp and mud carp. Mechanical lesions allow the bacteria (pseudomonas fluorescens) to invade the fish. Symptoms include inflammation bleeding from the skin and a loss of scales particularly on the sides of the abdomen. Blood shot on the fin base, necrosis of the terminal of fins and red blotches around the upper and lower jaws. Occasionally congestion and inflammation along the intestines also occurs.

Esomus Genus of south Asian fishes pertaining to the family of carp like fishes of cyprinidae. The body is elongated, strongly compressed laterally, the dorsal contour nearly straight. The mouth is small its opening slightly obliquely upwards. One pair of short and one of long barbels. The genus is distributed over India and Indochina.

***Esomus danricus* (Ham)** The flying barb. Besides the description of genus D-8, A-8, length up to five inches.

The colour is brownish with a faded red brown horizontal bands, which disappear with progressing age and is rimmed on the upper side with a lighter colour. It is distributed in small standing or flowing waters of all kinds also in rice paddies in India, Ceylon, Malaya.

Estrus After injection of spawning agent to fishes it will begin chasing each other excitedly. This phenomenon is known as estrus.

Estuary The area near the mouth of a river where sea water and freshwater meet. Here the freshwater from the river system dilutes the salt water and tides also push salt water up river to some extent depending on the strength of the tide, geologic formations the off shore currents and the quantity of freshwater entering the system via the river.

Ethology Science of behaviour.

Etiology Assignment or study of the causes of a disease.

Etroplus maculatus The body is high, strongly compressed laterally the fore head contour rising steeply, dorsal and anal fins are long and rather low. The body height increases with age . The large and active eyes are slightly oval. D-XVII-XX/8-10. The caudal fin is fan shaped slightly bent inward on the free edge, the mouth opening protrudes slightly in the shape of a snout. Length upto 3 inches, it is found in the fresh and brackish water of India and Ceylon.

***Etroplus suratensis* (B)** Body high, head relatively small, D XVIII-XIX/11-12 length up to 16 inches. It is commonly found in brackish and salt water in India and Ceylon.

Euglena It is a unicellular microscopic cell, slender with blunt anterior end and pointed posteriorly. Single flagellum at anterior end chloroplast in lobules.

Euryhaline The term referring to the capability of an organism to survive in environment of variable salinity.

Eurythermal The term referring to the capacity of some organisms to survive of wide range of temperatures.

Eurytopic Applied to organisms with a widespread distribution.

Eutrophication The process of over fertilization of a water body by nutrients that produce more organic matter than the self Purification processes can over come i.e. whenever the rate of synthesis and inputs of organic matter exceeds the recycling and output.

Eutropic An aquatic habitual with abundant organic matter and high level of biological productivity.

Evaporation Pond A body of seawater enclosed for natural evaporation and used to produce sea salt, traditionally a pure material . Much sea salt is produced in the mediter ranean where marine pollution has reached its purity.

Exfoliation The slughing of cells from the tissue surface.

Exo Skeleton The external shell or covering of a crustaceans.

Exocytosis The discharge from a cell of particles that are too large to diffuse through the cell memberane.

Exophthalmus Protrusion of the eye

Exotic Of Foreign origin or character. Anything exotic as plant, animals etc.

Exploitable Biomass The biomass of fishable size fish in a given stock.

Exponential Growth A type of growth in which the rate of increase in numbers at a given time is proportional to the number of individuals present. Thus when the population is small, multiplication is slow but as the population gets larger the rate of multiplication also increases. An exponential growth curve starts off slowly and increases faster and faster as times goes by. However at some points factors such as lack of nutrients accumulated wastes etc. limit further increase, when the curve of number against time begins to level off. The total curve is thus sigmoid.

Extension Service Worker Organisation or person forming the vital two way link between the farm and research organisation.

Extensive Pond Culture In this system the pond depends solely upon soil and water chemistry (and thus the available plant nutrients) and location (sun light, temperature) these factors limit the amount of primary production which takes place in the ponds and this in turn controls the rest of the pond system. Thus there is a good correlation between primary production and fish yield.

Extinct Species Any species no longer living on earth.

Extinction of Light When light penetrates into water it gradually dies out, it is extinguisdhed because the absorption and scattering. The rate of extinction is not the same for all the wave lengths.

Extracellular Situated or occuring outside a cell

Exuvium The cast (s) of the crustacean.

Exvessel prices The price paid to commercial harvesters for their catch when it is delivered to the dock.

F

FADS Fish Attraction devices.

Faecal Streptococcus A group of bacteria which are normally abundant in the intestinal tracts of warm blooded animals other than man, and are indicators of the contamination of water by the faeces of these animals.

Faeces Solid or semisolid material consisting of undigested food, bacteria, mucus bole and other secretion that is expelled from the alimentary canal, through the anus.

FAO Food and Agriculture Organisation.

Farm Dam Man made reservoir providing water for cattle or sheep in Australia, sometimes stocked with fish.

Fathom An old measure of water depth, about 1.83m (6 feet). It is supposed to be the length from left middle finger to right middle finger of a large man standing with his arm stretched.

Fatty Fish Fish when alive contains more than 2 per cent fat or oil in the flesh.

Feasibility Study Comprehensive evaluation of a proposal to farm prior to making an investment decision.

Fecundation The process of impregnation of fertilization of the egg.

Fecundity The number of eggs being ready for next spawning by a female fish. The different fish species have different fecundity e.g.

Catla catla - 80,000-1,00000 eggs/kg of Body weight.

Labeo rohita - 200,000-250,000 eggs/kg.

Cirrihinus mrigala - 125,000-150,000 eggs/kg.

Ctenopharyngodon idella - 52,000-134,000 / kg.

Hypopthal micthys molitrix - 120,000-170,000 / kg.

Cyprinus carpio - 132,000-160,000 / kg.

Feed Pond In some cases natural supplementary food is produced in separate ponds that are called feed pond.

Feminisation Modification of gender of fish using a hormone, drug or genetic manipulation.

Fermentation A chemical reaction produced by microorganisms (moulds, bacteria, or yeasts). A common example is the formation of ethanol from sugars.

$$C_6H_{12}O_6 = 2C_2H_5OH + 2CO_2$$

Fermented Fish Sauce Liquid made through fermenting of whole fish by their own enzymes (e.g. the gastric juices) and by certain micro-organisms in presence of salt.

Fertilizer Any organic or inorganic material of natural or synthetic origin which is added to a soil/water to supply certain elements essential to the growth of plants, or phytoplankton etc. in water.

Fibre The basic material used in the fabrication of yarns.

(i) Man Made fibre : Fibre manufactured by chemical or synthetic process.

(ii) Natural fibre : Fibre obtained in the natural form plant or animal Sources

(iii) Staple fibre : Any fibre of average length discontinuous fibres of regular length (cut man made fibres) or irregular length (natural fibres).

Fibroblasts A type of connective tissue cell which is responsible for synthesis of collagen.

Fibrosis The proliferation of connective tissues which consists of a high proportion of fibroblasts.

Field Capacity The point at which the soil contains all the water it hold by capillary and chemical attraction. Any more water added to soil at field capacity would drain away by gravity.

Figure of Eight Knot It is a knot made in the end of a rope by passing the end of the rope over and round the standing part, over its own part and through the bight.

Filament A fibre of indefinite length.

Filament Yarn Yarn filament.

Filamentous Bacterial Disease All stages of *P. monodon, P. indicus* and *P. merguiensis* being infected with *Leucothrix sp* show presence of fine colourless thread like growth on the body surface and gills under microscope. Respiration, moulting are interfered resulting mortalities due to the direct and indirect effects of the bacteria.

Filiform The inter-branchial septum is reduced and gill lamellae hang out like filaments and protrude freely into the branchial chamber. Such gill is found in bony fishes and is known as filiform,

Filinia Belongs to the phylum Trochelminthes order monogononta, family Brachionidae. They have 2-3 very long moveable appendages extends from anterior side, body shape indefinite, foot absent.

Fillet Strips of flesh out parallel to the central bone of the fish and from which fins, main bones and sometimes belly flap have been removed, presented without or with skin.

(i) Single fillet or side is the flesh removed from one side of a round fish e.g. cod two such fillets are obtained from each fish, the bellywall through trimmed is sometimes left on turn side particularly used for single fillets of salmon.

(ii) Full nape fillet : a single fillet which includes belly flap and often rib bones commercial or quarter nape fillet belly flap removed essentially boneless.

(iii) Block fillet is the flesh cut from both sides of a same fish, the two pieces remaining joined together along the back, usually made from smaller size of fish also called angle fillet (for haddock in U.K.) and butterfly fillet, may also be called cutlet.

Filter Biological Water treatment system in which there is a large surface area occupied by micro organisms that oxidise dissolve organic matter, ammonia and nitrite to less harmful products.

Filter Feeders The animals whose feeding mechanisms are a filter. It means creating a current running through the filter a scraping device to move food from the filter and water exit. The filter's function may be accompanied by a mucus net put out by the animal and ingested along with any thing it has trapped. The filter itself may be made of cilia, setae antenna or antennules and may be part of gill or maxilliped. The cilia or setae may beat in concerted waves to create a current which moves detritus or plankton part to the net. Either the cilia or setae or maxllipeds than deliver the item to the mouth the filter will absorb particles of an acceptable size and reject those that are too large.

Filter Mechanical A water treatment system which mechanically strains or collects suspended particulate material from the water.

Fimbriated Fringed or with short fimbriate or processes.

Fin rot Caused by Aeromonas. The fins become opaque at the margins and the opacity usually progress toward the base. Fins also become thicker because of proliferation of epithelium and in advanced cases are frayed with rays protruding like stretched fingers. In advanced tail rot the whole caudal fin may be lost, followed by gradual erosion of the peduncle.

Fin spine It is a stiff, unbranched support of fin. A fin-spine may be simple, jointed or serrated. The fins supported by both spines and rays, the spines are anterior to the rays.

Fin-ray Fin-ray is a branched or forked stiff structures supporting and strengthening the fin membrane. In primitively evolved fish, only fin rays are present, there being no fin spines.

Financial Risks Risks due to government financial policies use and dependences on government policy instruments, terms of credit, changes in operational costs.

Fineness General term for the size or the dimension of textile products measured as weight per unit length, length per unit weight, cross sectional area or diameter.

Fingerlings A fish that is about the length of a man's finger (6-10 cm).

Fins Flattened organs usually supported by fin rays used in Swimming. Most fish have proved pectoral and pelvic fins; homologous to the forelimbs and hind limbs respectively of terrestrial vertebrates and used for controlling the angle of ascent or descent. In some species the Pelvic fins are small and modified for compulation. Some fishes also have one or more dorsal fins and one or more ventral fins (anal fins) used for preventing sideways movement and rolling and a caudal fin (tail fin) used for propelling the fish in continuous forward motion.

Fish The word fish is derived from the French word "fiche" which means "means of fixing". In scientific jargon, fishes are known as pisces. Pisces is a latin word. Fishes are cold-blooded (or poikilo-thermous *i.e.* their body temperature varies with that of the environment) aquatic Vertebrates which breathe by means of pharyngeal gills, propelling and balancing themselves by means of fins. Fishes are the oldest verteberates evolved in silurian and Devonian periods about 350-450 million years ago. Fishes are the most numerous of the vertebrates comprising 42.6% of all the vertebrates. Fishes come in all sizes ranging from the tiniest which is 10 mm in length and weighs 1.5 grams, to a gigantic 20 meters weighing 4000 kg. There are about 25,000 fish species inhabiting all types of ecosystems of the world. India contains about 1479 species of fish.

Fish Flakes Product prepared by headed and gutted white fish such as cod and haddock washed, brined and steamed, the bones removed and the cooked flesh broken up into flakes and canned.

Fish Furunculosis It is caused by *Aeromonas salmonicida*, a non motile gram negative bacterium. Boil like lesions are the first symptoms, other symptoms are blood shot fins, bloody discharge from the vent, many small hemorrhages in muscles and other tissue and necrosis of the kidney.

Fish glue Fish glue is prepared from different wastes, e.g. bones, scales and fin etc. discarded during processing. These are washed, ground and cooked with acetic acid in steam, liquid is separated and condensed to form the fish glue, which is used as an adhesive for several purposes.

Fish Growth Rate This is calculated as

$$\frac{\text{Total weight gain of certain fish (gram)}}{\text{No. of fingerlings and their culture days.}}$$

Fish in Jelly Pieces of fish or minced fish cooked heated in acidified brine or with vinegar or fried or smoked and packed in gelatin or gelatin and pectin or aspic (dissolved by heat) in Germany usually with cucumbers, onions, and spices and must contain 50% fish.

Fish Kill The sudden mortality in a fish stock; usually caused by lack of oxygen or poisoning never by disease.

Fish Liver It is used for preparation of medicinal products (high vitamin A & D) and as food. Liver of lower vitamin such as rays, spotted dog fish can be used for industrial products.

Fish Meal Fish & Fish processing offal dried often after cooking and pressing (for fatty fish) and ground to give a dry, easily stored product that is a valuable ingredient of animal feeding stuffs.

Fish Mortality The fraction of an exploitable part of a stock or of an age of year class which is removed as catch within a year.

Fish Oils Oils of the drying and semidrying types extracted from all of the body of fish (Fish liver oils from liver only); They are extracted mainly from fatty fish e.g. Herring, the oil content is mainly in the body and not in the liver, may also be extracted from fish waste. Fish oils are used in the manufacture of edible fats, soaps, and paints for leather, dressing and lnoleum manufacture.

Fish Paste/Fish Spread Fish mixed with or without spices or other flavouring ingredients, ground to fine consistency of lowered moisture content often with added fat used as sandwich spread.

Fish Pickles It is prepared by cutting the edible portion of the fish into small pieces, followed by deep frying in vegetable oil and are subsequently mixed with vinegar and salt for preservation, along with fried condiments and species for flavour development. The material is then generally kept for a minimum of 24 hrs for maturing before packing. This is a traditional product of the country and is now gaining popularity in middle east countries.

Fish Pie Fish often minced and sometimes mixed with vegetables particularly potato and baked may have pastry casing.

Fish Portion A piece of fish cut to reasonable size for the individual for retail sale, may be all or part of a Fillet, Steak, may be fresh, fried or frozen.

Fish Protein Concentrate (FPC) Any stable fish preparation intented for human consumption in which the protein is more concentrated than in the original fish, marketed in granular, powdered or solvent extraction. Two type exist i.e. A and B of which type A is odourless and colourless while type B not principally due to different fat content (less than 1% in type A, less than 10% in type B).

Fish Pump It is a device used to pump the fish from purse seine net on to the vessel or transport fish around farms through wide bore tubes causes less physical damage than netting.

Fish Salad Cooked salted or marinated fish diced with spices, diced onions, cucumbers, and vegetables mentioned for the type of salad mixed with vinegar and edible oil or mayonnaise, minimum fish contents are fixed for various special types.

Fish Sausage Fish flesh ground with a small amount of fat seasoning and sometimes a cereal filler, packed into sausage casing. Sometimes cooked, the contents may be smoked before filling the case, or the whole sausage smoked afterwards, may be sold skinless or with skin on. Tuna meat is much used for sausage manufacture.

Fish seed Fertilized eggs, fry or fingerlings of fish used for growing in fish culture.

Fish Silage Liquefied fish or fish waste produced as a result of self digestion after the addition of acid or as a result of fermentation of the waste mixed with molasses and yeast. The liquid silage can be concentrated to bring the water content down from about 80% to 50% used for animal feeding usually in areas close to the point of manufacture.

Fish Soup Soup made from fish or other marine seasoning added and eventually served with flavouring vegetables, sometimes containing pieces of fish.

Fish Stearin A solid produced by separating chilled fish oils, used mainly for manufacture of lubricants and low grade soaps.

Fish Waste All parts of the fish discarded during processing for human consumption also called fish scrap, Fish offal. It is used for manufacture of fish meal and oil for the feeding to pets, fur-bearing animals and hatchery fish and for the manufacture of a variety of by products including, pearl essence, isinglass, glue proteins, vitamins.

Fisheries Conservation Those measures concerned with the protection and preservation of fish and other aquatic life.

Fisherman's Knot This knot is made of two overhand knots which slide together when the ropes are pulled forming a firm knot.

Fishery A fishery is the complex of interactions within and between the population (s) of fish being harvested, the population (s) of fishermen, and the environments of each. Fisheries primarily started as capture fisheries. Fish culture on commercial scales is a recent innovation to fisheries.

Fishery Science It is a multidisciplinary subject consist of several basic sciences. Fishery Science is the application of scientific knowledge concerning fish populations to the problems of obtaining the optimum production of fishery products, whether stated in tons of factory material or in hours of angling pleasure.

Fishing Method Means the manner or method in which fish may be captured.

Fixation of Tannin Treating the tanned net in a solution of some chemicals like potassium bicarbonate etc. to make tannin in solution in water so that it will stay longer in the twines and ropes.

Fixed Cost/Average Fixed Cost Cost of production which do not vary with the level of output. Average fixed cost refers to fixed cost per unit of output.

Fixing a Net Fixing a piece of net to a line or rope around all or some of its edges at any definite degree of closeness or openness of the meshes. Also called mounting or hanging a net.

Flag Piece of bunting surmounting to aid visual location.

Flag Pole It is a bamboo pole of about 4-4.5m in length and used to mark the position of two long line gear. It is kept in a upright position in water by tying the top end of the float line to the basis of the flag pole and attaching the glass float about 1.5 m above the base pole.

Flake A soft form of ice used in processing delicate crustacean flesh.

Flashing A sudden abnormal movement of fish in which they turn on their flanks, usually due to an external parasites.

Flat Knot Sea knot.

Flax A best fibre obtained from the plant *Linum usitatissimum*. Grown in USSR and west Europe. Yarn and clothes made from Flax are called Linen.

Fleet End Rope A piece of rope folded and interlocked to the central part of the bridles of the distal end of gill net fleet forming two arms one of which bears the master float and the other Master sinker.

Flexural Stiffness This is the resistance of a specimen to lateral or bending deformation.

Float Line The rope on which floats are fastened. It is attached to the top edges of a net. Also called cork line etc.

Flock Fragments of bacterial or other biological growths sloughed off from the surfaces of biological filter and Aquaculture tanks.

Flood A rise of the sea, river or stream from its normal height. Generally it is due to storm or volcanic action or heavy snow fall or excessive rainfall.

Flood Frequency The average interval of time between floods equal to or greater than a specified discharge or stage, generally expressed in years.

Flood Plain Relatively level part of a river valley adjacent to the river channel, formed from sediments deposited by the river during periods of flooding.

Flood Way The channel of a river or stream and those parts of the flood plains adjoining the channel that are reasonably required to carry and discharge the flood water or flood flow of any river or stream.

Flume A specially designed and shaped channel used for conveying water by gravity, normally brick or concrete lined for fast flowing water.

Flushing rate The rate of water exchange in an aquatic system. Lakes with high flushing rates are those that relatively high rates of inflow and outflow perunit time in comparision to the volume of water that is held by the lake.

Flying Food Edible animals generally air borne insects and other arthropods which live close to the water surface and are caught by the fishes by means of jumps or when falling directly upon the surface of water.

Follicle Stimulating Hormone (FSH) A Gonadotropin produced by the anterior pituitary gland. It acts on the ovary to stimulate the growth and maturation of the tissues forming follicles and ova; which under the action of luteinizing hormone, mature and are released from the ovary. It also stiumlates spermatogenesis in males.

Food Chain The transfer of food energy from the source of plants through a series of organisms with repeated stages of eating and being eaten.

Food Conversion Efficiency (FCE) The amount of food necessary to produce one unit weight (wet) of animals. For example if a pond produce 1250 kg of prawns and 3200 kg of food were used during the rearing period the FCE is

$$\frac{3200}{1250} = 2 \cdot 56$$

It follows therefore that the lower the FCE is the better the efficiency (of conversion into final product) of the food is. The FCE of wet feeds be much higher than that of dry feeds because of the difference in moisture content. FCE is the inverse of FCR

$$FCE = \frac{\text{total fish body weight gained}}{\text{total weight of feed intake}}$$

Food Conversion Factor This indicates the conversion of food into the flesh. It can be calculated as,

$$\frac{\text{Consumption of a certain feed (kg)}}{\text{gross increment of fish given this feed (kg)}}$$

Food Conversion Ratio (FCR) As for food conversion efficiency except that it is written as a ratio i.e. food conversion efficiency of 2.8 is written as a food converston ratio of 2.8:1. This means that 2.8 kg of food is necessary to produce 1 kg of prawns (live weight). It can be calculated as

$$\frac{\text{Food consumed (net weight)}}{\text{Fish body wt. (net) gain}}$$

or

$$\frac{\text{Food applied}}{\text{Total weight at harvest} - \text{total weight at stocking}}$$

Food Efficiency/Food Conversion Rate This gives the percentage of food conversion in fish flesh. It may be calculated as

$$= \frac{\text{total production of animal} \times 100}{\text{total consumption of feed (kg.)}}$$

Food Utilisation Rate It refers mainly the uitlisation rate of crude protein in the diet. It can be calculated as

$$\frac{\text{Protein increase in Fish body}}{\text{Protein in consumed food}} \times 100$$

Food Webs The interrelationships and intersections of food chains found in a particular place at a particular time.

Foots Liquor remaining from steamed cod liver oil may be concentrated for use as animal food, manufacture as cod liver meal. Also liquor ramaining after cooked fish is pressed in fish meal manufacture.

Forage base The supply of food resources available to a particular species.

Forage fish Small sized fish generally utilized as food by larger predatory fish

Fore Hand Line A Three strand hemp or man made rope used in whale catching. It connects the harpoon to the fore hand line.

Fork Length Length of fish measured from the tip of snout to point of division of tail or caudal fin.

Formalin A mixture of 40% formaldehyde 8% methyl alcohol and 52% water. It is a powerful reducing agent and is used as a disinfectant, germicide, fungicide, and also as a general preserving solution. In contact with the skin formalin may cause irritant dermatitis and ingestion can cause severe abdominal pain.

Forum of Fisheries Professionals Established in 1991. The permanent address is plot 176, sector 12, MVP colony Visakhapatnam. The objective of the forum is to endeavour to pool up authentic professional knowledge on various aspects of Fisheries development. To organise professional opinion as a major force in the moulding of policies of the various governments in respect of fisheries development.

Fouling The assemblage of organisms that attach to and grow on underwater objects, also the deleterious accumulation of dissolved and particulate material in a body of water or pond bottom.

Fouling Organism Any aquatic organism with a sessile adult stage that attaches to and fouls underwater structure of ships.

Foundation Line A strong thick twine streached between two supports on which begins the making of a webbing.

Four Seam Trawl A trawl comprising four panels, top, bottom, and two sides.

Free Board Upper part of a canal, dike pond or similar structure between the water level and the top of the structure.

Freeze Drying A process of dehydration under vaccum starting from the frozen state. The frozen substance having been placed under reduced pressure. The existing ice is sublimated (directly transformed to vapour) leaving a fine porous structure favourable to rehydration.

Fresh Fish The fish is not or has never been frozen, cooked cured or otherwise preserved and that it exhibits a clean natural odour and physical characteristics that are representative of the species being in good conditions.

Frozen Fish Fish that has been subjected to freezing in a manner to preserve the inherent quality of the fish by reducing their average temperature to–18°C (0°F) or lower and which are then kept at a temperature of –18°C (0°F) or lower. The reduction of temperature in the fish has to be achieved in a sufficiently short time to give the product of best quality, this depends on the type of product. Regulation with regard to maximum time have been laid down in various countries e.g. in U.K. the temperature of the whole fish is to be lowered from 32° F to 23°F or lower in not more than two hours and the fish be kept in the freezer until the temperature of the warmest part of the fish has been reduced to -5°F or lower, according to U.S. recommendations the temperature should be lowered to 0°F in 36 hours or less.

Fry Young fish upto 1 inch (2.5 cm) long.

FSI Fishery Survey of India, is responsible for studying the fishery potential of the country. The head office is at Bombay.

Fucoxanthin A red brown pigment present in and characteristic of brown algae. It masks the chlorophyll in these plants.

Fungi Essentially land plants that have no chlorophyll. They live on dead plant tissue.

Furuncle The small area of infection in or just under the skin which develops into a raised abscess, furunculosis disease.

G

G-Hook Forged shaped hook to connect with a recessed link (coupling link).

Gamete A specific haploid cell produced during sexual reproduction that fuses with another gamete to produce a zygote from which the next generation develops. In general such fusion occurs between male gametes and female gametes. The female gametes are often sessile whereas male gametes are generally motile.

Gangions The Series of hooks and leaders attached to long line gear.

Ganoid scales They are found only in some of the present-day fishes but were common in a group of extinct fishes. A ganoid scale lies deeper in the dermis, the vascular layer has few pulp cavities and the upper most layer is composed of a hard, shiny, translucent material, called ganoin, Ganoid scales are generally rectangular or rehomboid and closely set in diagonal rows. They are articulated in characteristic "peg and socket". Palaeniscoid scale and Lepidostoid scale are the two types of ganoid scales. Palaeoniscoid scale is found in *Polypterus*. Lepidostoid scale is found in *Lepisosteus* . The epidermis overlying these scales may possess true denticles.

Garos Ezymatic preparation from hydrolysed livers of mackeral used in Greece to accelerate maturing of freshly salted Sardine and anchovy.

Garum Sauce produced in Mediterranean regions by mixing whole ungutted fish with concentrated brine and exposing the mixture in jars for long periods to the sun.

Gas bladder It is one of the characteristic feature of the true fishes. It is quite often regarded as swim bladder or air bladder and found to be highly developed in spiny rayed teleosts (Acanthopterygii). It is an accessory respiratory organ also helps in sound production and sound perception, fat storage. It is an important hydrostatic organ, contains a gas secreting complex, is composed of a gas gland covered with blood vessels. The gas bladder is completely absent in Echeneifoemes, Symbranchifoemes, Saccopharyngiformes, Gobeisociformes and elasmobranch fishes. The gas bladder may be oval, flusiform, tubular heart shaped, horse-shoe shaped or dumb-bell shaped.

Gas Bubble Disease Appearance of gas bubbles under the skin in blood vessels etc. which results in the tearing of tissues. Spreading of fins and eye disorders, which can result in death. This is usually caused by the blocking of the blood vessels with gas bubbles. This disease is caused by a sudden change in gas pressure. Such as when the total pressure of the dissolved oxygen in the blood is higher than the dissolved gases in the water which is being breathed.

Gas disease A fish disease caused by excess of nitrogen or oxygen resulting in accumulation of gas in body spaces.

Gastritis the inflammation of the stomach.

Gastroderm The lining of the gut.

Gastrolith/Stomach Stone Mass of calcium carbonate found each side of the cardiac stomach, especially in cray fish, forms reservoir of calcium for shell remineralization after moulting.

Gauge Mesh An implement used in the braiding of nets. Made of wood or some cheap metals. It regulates the size of the meshes of the webbing and ensures uniformity of mesh size.

Gear efficiency A measure of the proportion of a group of fishes encountering harvesting gear that are captured by it.

Gelatine Soluble protein that can be prepared from the skins and swim bladders of fish but largely produced from animals skins, used in the food and other industries.

Gene The unit that controls hereditary characteristics transmitted in the chromosomes.

Gene Bank A collection of cell cultures seeds frozen sperms or ova etc., kept as a means of holding representative genomes of any type of organism.

Gene Flow The introduction of new genes to a population from an outside source by interbreeding. Thus increasing the degree of genetic variables.

Gene Pool The total available genes within an interbreeding population at any given time.

Genotype Genetic make up of a fish, that portion which is inherited.

Geosmin A compound synthesized and secreted by the blue green algae and Actinomycetes. The flavours of the compound is earthy musty, which absorbed by the fish giving them a off flavour to make them unmarketable.

Gerris/Water Strider Insect belong to order Hemiptera family Gerridae. This insect lead a gregarious life, head small and body slender forelegs small. While two pairs of legs 2-3 times as long as the body, live on the surface film grow to 2-15 mm.

Gibbing The process of removing the gills, long gut and stomach from a fish such as the herring by inserting a knife at the gills. The milt or roe and some of the pyloric caeca are left in the fish.

Gill A respiratory organ of fish, molluscs, crustacean and some worms. A gill is constructed of finely divided featherlike tissue that is well supplied with blood. The gill is often enclosed in a supportive or protective structure made of bone shell or chitin. In teleost fish the gills are covered by an operculum. The finely divided structure of the gill makes possible maximum contact of the blood in the gill with surounding medium. The blood brings carbondioxide to the gill and this gas is passed from the blood to the surrounding water. At the same time oxygen diffuses from the water in to the gill and into the circulatory system. In many organism the water is forced over the gills by muscle action respiration in gilled organisms is not a passive process.

Gill Bar One of a number of skeletal structures in the wall of the pharynx in fish that supports the tissue separating successive gill slits. They bear numerous fine filament and are richly supplied with blood into which oxygen diffuses from the water. Gill bars are also present in lower chordates.

Gill Books Respiratory organs of the king crabs, consisting of groups of thin plates attached to the abdominal appendages. They contain circulating blood. Their structure is similar to the lung books of terrestrial arachnids indicating an evolutionary relationship between these groups.

Gill Cleft An opening through which water leaves the body of a fish after passing through the gills.

Gill Cover The fold of skin providing external protection for the gills apparatus of most fishes, it may be stiffened by bony plates and covered with scales.

Gill filaments The epithlial layer of the septum forms a series of delicate, thin-walled parallel evaginations on either side of the septa and are called the gill-filaments or gill lamellae. The gill fillaments are richly supplied with blood capillaries when water passes through the gill-clefts over the gill-fillaments, exchange of gases takes place.

Gill Fungus / Gill Rot Caused by Branchiomyces sp. in cultivated fish in ponds having abundant decaying organic matter.

Gill membrane The thin layer covering the gills and permitting exchange of gases for respiratory purposes.

Gill Net A rectangular wall of webbing provided with a head rope at the upper edge bearing floats and with or without a foot rope at the lower edge bearing sinkers. The mechanism of capture being enmeshing of fish either in a single lumen or several lumens of the meshes at the gill region or any part of the body between the gill region and dorsal fin.

Gill Pouch One of several pairs of compartments lying laterally to the respiratory tube in cyclostomes. They contain gill filaments.

Gill rakers Epi-and ceratobranchials bear dorsally a double row of small bony processes, called gill raekers. Gill rakers protect the tender gill filaments from abrasion by ingested materials that are coarse in texture. Gill rakers are also specialized in relation to food and feeding habits of fishes. In Omnivores, they are very stubely and unadorned. In many plankton feeders, the gill rakers are elongated, numerous, and variously lamellated or ornamented, to augument efficiency in straining. In carnivores and predators, the gill rakers are widely spaced and in the form of hard rasping organs.

Gill rot A disease commonly affecting the gill.

Gill slits These are narrow openings-usually five pairs in sharks and rays, from which water is expelled while breathing. In other elasmobranch fishes also, the usual number of functional gill slits is five only in *Heptanchus* and *Hexanthus* the number is 7 and 6 pairs respectively. In the chimaeras, only 4 gill-slits are there. Position of gill-slits is a factor in the classification of modern cartilaginous fishes (Euselachii) which are divided into two sub orders : 1. Pleurotremata, having gill-slits on the lateral sides and the spiracle is either closed or bears a pseudobranch, and 2. Hypotremata, having gill-slits on the ventral side, the spiracle in them serves to draw the respiratory current and is provided with valves.

Glazing The process of protecting unwrapped frozen fish against drying in cold storage by dipping in cold water or by brushing or spraying with cold water immediately after freezing to form a thin protective skin of ice.

Gnathostomata Gnathostomes possess true jaws, paired appendages well-formed endoskeleton and back-bone more advanced brain and sense organs. They are graded according to increasing complexity of organization into Pisces, Amphibia, Reptilia, Aves and Mammalia.

Gonad Organ in which either sperm or eggs are produced general name for either the testis or the ovary.

Stages of gonadal maturation of carps

Stages of maturation		Condition of gonad: Female (Ovary)	Condition of gonad: Male (Testis)
(i)	Immature	A transparent long narrow strip	Transparent thread like
(ii)	Immature	Thickened strip trans-lucent.	Translucent thread like
(iii)	Maturing	Opaque granular some-what greyish occupying about 1/3 of the body cavity.	opaque pinkish whitish thin strip like.
(iv)	Maturing	Dull greyish in colour ova discernible like granules. occupying about 1/2 of the body cavity.	Thick strip milky oozing whitish fluid on applying pressure on belly.
(v)	Mature	Dull grey to greenish occu-pying atmost the entire body cavity, ova distinctly round in shape.	Thickened band like oozing whitish fluid on applying pressure on belly.
(vi)	Spawning	Having loose eggs in ovarian wall, eggs oozing through genital aperturre on applying pressure on belly.	oozing milt freely on applying pressure on belly.
(vii)	Spent	Ovary blood shot pinkish brown mass but greatly shunken.	oozing milt freely on apply-ing pressure on belly.

Gonadal Somatic index It is obtained by weight of gonad x 100 ÷ body weight.

Gonadotropic Hormones Hormones of vertebrates secreted by the pituitary gland which control the activity of the gonads including their production of hormones.

Gonochoristic Species Those species in which any individual spends its entire adult life as either a male or female but not as both.

Gonopodium Mating organ of the live bearing tooth fishes developed on the male fishes by a transformation of the anal fin and which serves to bring near to or introduce into the female sexual opening the spermatozoa which generally are joined into bundles.

Gonopore The external opening of the reproductive system of a crustacean.

Gossypol A compound of cotton seed cake which is deterimental to livestock.

Grab Line A line becketted round the outside of life boats just below the gunwale forming loops to which men in the water cling on.

Gradient The steepness of a sloping surface which may be expressed either by the angle in degree from the horizontal or by representing the vertical distance and the horizontal distance as a ratio.

Graft A piece of transplanted tissue.

Gram's Stain A stain containing crystal violet and safranin used for bacteria and the basis for the division into gram positive and gram negative bacteria. The former retain the deep purple colour of crystal violet and the gram negative bacteria stained red with safranin.

Granton Trawl A type of otter trawl used for deep sea trawling.

Granuloma A lesion resembling a tumour which results from chronic inflammation and consisting primarily of aggregation of macrophages, epitheloid cells and some connective tissue elements.

Granulomatous inflammation The type of inflammation characterized by the exudate composed of macrophages epithelioid and gaint cells and by connective tissue formation.

Grapnel Three or four metal hooks set together back to back; may be used as a small anchor, or with a line attached it is used for picking upon grapping for ropes or other items.

Gravlax Fillets of salmon rubbed in with a mixture of coarse salt, sugar and white pepper meat side against meat side with dill, and pressed in a chilly place, considered to be best after 24 hours.

Gray Whale A relatively small (about 15 m or 50 foot) long-slender baleen whale. It is usually gray to gray black and spotted with white. The grays are found most often closer to shore as opposed to out in the open ocean.

Green Fish Salted white fish that having been stacked for two or three days to press out as much pickle as possible are ready for drying also called green cure, green salted fish. In Newzealand the term refers to fish caught and stored temporarily (fresh chilled or frozen) in a unprocessed form.

Green House Effect A blanket, term used for describing a world wide heating trend linked with a general rise in the level of atmospheric carbondioxide.

Green Manure Wild grasses and plants used as manure are called green. These manure rot and decompose easily providing ideal environment for bacterial propagation. They are good for application in fish ponds. The nutritional elements varies from plant to plant.

Gregarian Disease Protozoans parasites commonly found in the digestive tract of larvae, post larvae, juveniles and adults of *P. monodon* and be detected microscopically. Large number of this protozoans which utilize a mollusc species as intermediate host could interfere with particle filtration through the gut or the hepatopancreatic duct.

Gregarious Living together or swimming about in shoals.

Groin A man made rocky structure built perpendicular to a bench to inhibit the movement of beach material and to prevent beach erosion.

Ground Water Water that occupies pores and crevices in rock and soil below the surface and above a layer of impermeable material as opposed to surface water, which remains at or close to the land surface. The upper limit of the ground water is the water table, whose level varies according to the quantity of water entering the ground water compared to the quantity lost.

Growth Hormone/Somatotropin Pituitary hormone which controls growth and repair of tissues.

Guanin/Guanin Crystals An excretion product that is deposited as a Shining metallic crystalline layer in the skin, iris, swiming bladder and some other parts of the body of fishes. The crystals have a strong refractory action of light and produce the silvery shine on fishes as well as the glow colours when combined with other aliments.

Guild A group of species in a fish community that exploit the same resources in a similar manner.

Gull A fish eating bird of the sea coast and river estuaries

Gut The alimentary canal through which food passes in digestion, assimilation and excretion.

Gyre Circular ocean currents created by surface winds, rotation of the earth and the corioils effects.

Gyomiso Fermented fish paste prepared from mixture of macerated fish flesh, salt and wheat bran inoculated with a fungus called *Aspergillus oryzae* formerly produced in industrial scale.

H

Habitat The space occupied by an individual or group of individuals including every thing in that space both living and nonliving.

HACCP Hazard analysis critical control point programme involves description of the product in relation to its intended use, flow diagrams for manufacture and distribution, evaluation of risk analysis and determination of critical control point on the flow diagram.

Hadal Zone The term used for the ecological zone at ocean depth greater than 6 km (20,000 ft.). There is no light at this depth the water pressure is over 600 atmosphere can be as high as 1200 atmospheres in the ocean trenches. Both the temperature and salinity of hydal zones are fairly constant. The salt content is about 3.47%. The bottom itself is composed of red clay of terrigenous origin and radiation ooze. The dominant animals are omnivorous, crustaceans, echinoderms polychaetes and molluscs.

Haematophagous Feeding on blood or lymph.

Haemorrhage Internal bleeding and subsequent clotting caused by the rupture of blood vassels.

Hair weed The genus of long and filamentous green algae *e.g.* spirogyra.

Half Salted Fish Fish removed from brine before it is fully cured also known half fresh fish.

Halocline A layer of water in which the salt content changes rapidly. It forms a boundary between regions of different salt content.

Halver Product of single bar cut leaving three bars to one knot.

Halving Becket Ring Metal or plastic rings fixed around cod end through which the halving becket passes.

Hamburg Plankton Net The Humbarg plankton net is designed to for collection of larger quantites of plankton for instance necessary for the detection of tracer elements for radionuclides. The net is equiped with a bag net of 500 micron mesh size. The overall length of the net is 10 meters. When the Hamburg plankton net is towed the drag to the fine meshed net is supported by a coarse meshed covering net and 5 stainless steel rope along the net. A netsode can be attached to the depressor so that the exact depth of net is indicated abroad. For security of the net a predetermined breaking point has been constructed into the bridle; which makes the net collapse when the dynamic pressure becomes too strong.

Hand Line The lines operated with few hooks by a singleman paying constant attention.

Handi A wide mouthed vassel made of mud or tin used for transporting the fish seed.

Hanging Coefficient It is the hanging ratio expressed in decimal fraction and used more commonly to express hanging.

Hanging Ratio It is the ratio between the length of rope and the length of the netting.

Hapa A miniature tank like structure made of cheap cloth and fixed to poles in water for temporary storage of fish seed, and fish breeding.

Hapatomegaly The enlargement of the liver.

Haploid Having half the normal number of chromosomes most gametes are haploid so that when fertilization occurs the diploid condition is maintained.

Hard Cure White fish particularly cod that have been dry salted. Dried to a moisture content of 40% or less, also called full cured fish, full pickle fish.

Hard Fibre Fibres obtained from the leaves or leaf stalks of various monocotyledonous plant like sisal and abaca.

Hard Smoked Fish Fish subjected to prolonged period of cold smoke unit e.g. golden cure.

Hard Water Natural water which contains 40 mg/litre or more total alkalinity as equivalent to $CaCO_3$ are hard water. Hard water is more productive than soft water.

Hardening of eggs The eggs fallen into the cavity of ovary do not have definite shape egg shell is soft and tightly surrounds the cell 'Kernel' which includes the nucleus and the mass of yolk. When the eggs falls into water it assumes a round shape and within a short time begins to swell. The water penetrates between the shell and kernel of eggs and thus Perivitalline space develops, the micropyle closes, the swelling usually lasts for about 1-2 hours. The eggs then takes its final forms and becomes water hardened.

Harpon Line A line attached to the harpon used for catching whales or big fish.

Hatching The process of brusting of the egg membrane and liberation of the baby fish.

Hatching Hapa Hatching hapa consist of two separate pieces of hapa—the outer hapa and inner hapa. The inner hapa is smaller and fitted inside the outer hapa. The outer hapa is made up of thin markin cloth or nylon stitched in the shape of rectangular box without any cover and have ropes on upper and lower end; The mesh size of outer hapa is smaller whereas the mesh size of inner hapa is bigger. The standard size of the outer hapa usually used for the purpose is $2 \times 1 \times 1$ m. The inner hapa is made of round meshed mosquito net cloth with the dimension of $1.75 \times 0.75 \times 0.5$m. It has also arrangements of tying in the outer hapa.

Hatchling Young one just hatched out of the eggs.

Hawser A three strand rope generally over 127 mm in circumference of flexible construction used for towing, mooring, warping and securing.

Head Loss The loss of head occurring through friction, change of speed etc. as water is pushed through a pipe or other hydraulic structure.

Head Rope The rope on line to which the upper edge of the net is finely attached. Floats are attached to this rope for keeping net a float.

Head Setting Process of stabilizing the fibres by subjecting them to heat. This improves its resistance to ceasing.

Head The level at which water is held or can rise to allowing it to flow to lower levels, push through pipes etc. also pumping head, the level to which a pump can push water.

Head Water The portion of a river basin that occurs at the heighest elevations for the water shed, the source of a river system.

Heavy Salted Fish Fish cured by adding salt so that product contains approximately 40% salt on dry weight basis.

Heavy Salted Soft Cure Product obtained by heavy salting but without hard drying, so that salt content averages about 17% on weight basis and moisture content 47%.

Hectare Metric unit of area 10,000 square meter or 2.46 acres.

Hemibranch (demibranch) The gill filaments generally develop on both sides of the interbranchial septum but in some instances they are present only on one side. The series of gill-lamellae on one side of the septum constitutes only half a gill and is thus called a hemibranch or demibranch.

Hemorrhagic Septicemia It is caused by *Aeromonas liquefaciens*.

Hemp Fibre obtained from the bark of the hemp plant *Cannabis saliva* grown in Italy and USSR.

Heparin A substance that prevents blood clotting by neutralizing prothrombin and stopping the action of thrombin. It is present in tissues of mammals and is secreted by some blood sucking animals.

Hepatopancreas A part of the digestive tract in crustacean that extends from the midgut and is composed of numerous ducts and blind tubules.

Hepatopancreatic Parvo-live Virus, (HPV) Disease Retarded growth loss of appetite and occasional white opaque areas on the abdominal muscles are noticed in juveniles and adult of *P. monodon* and *P. merguiensis.* Fouling of exoskeleton occur by benthic diatoms protozoans and filamentous bacteria. The virus cause cell death and shrinkage of the hepatopancrease leading to abnormal metabolism and death.

Hereditary Traits The sum total of the all possibilities of development which an organism transmits hereditarily to its progeny.

Hermaphrodite A species capable of producing male and female sex cells either synchronously or by changing from one sex to the other.

Heterocercal caudal fin This type of fin is found in condrichthyes and some primitive bony fishes. In the heterocercal type, the vertebral column bends upwards and thus divides tail into unequal parts. The hypochordal lobe is much larger than the epichordal lobe. The tail is therefore asymmetrical.

Heterogeneous Growth Different rates of growth occurring in the same population of animals leading to a wide range of sizes at the time of harvesting.

***Heteropneustes fossilis* (Bloch)** Belongs to family Clarride with elongated eel like body which is flat at the head but strongly compressed towards the rear. The wide mouth with thick lips and short pointed teeth on jaws and palate carries four pair of barbels. The skin is scaleless and shiny, the eyes are small, D 6-7, A 60-79. The fleshy dorsal fin is only indicates by a fold in the skin. The long anal fin is slightly separated from the caudal one, length upto 28 inches.

Heterosis Hybrid or interstrain vigour produced by crossing or cross-fertilising different species strains or stocks.

Heterotrophic Organism Organisms that are dependent on organic matter for food.

Heteroxenosu parasite A parasite that requires more than one host to complete its life cycle.

Hetrodont Hetrodont dentition is found in some sharks such as in Heterodontas, where the anterior teeth are recurved and posterior teeth are plate-like with scroll-like surfaces for crushing.

Histo phagous Feeding on tissues of animals

Histolysis The breakdown of host tissues.

Hold fast The attachment organ of some parasites

Holobranch A complete gill is known as holobranch. Each halobranch consists of an anterior (Oral) and a posterior (aboral) hemibranch.

Homeostatis It refers the maintenance of physicological conditions within narrow limits of an organism.

Homeotherms Warm blooded animals, birds and mammals. They have much greater flexibility in selecting their environment but more insulation and much more food than do the Poikilotherms (cold blooded).

Homocercal caudal fin It is mostly found in higher bony fishes. In it, the fin is symmetrical externally but asymmetrical internally. The hypochordal and the epichordal lobes appear similar but actually the ventral lobe is greatly enlarged and divided into equal lobes. The posterior end of vertebral column is turned upwards and greatly reduced.

Homodont Possessing teeth which are all similar as in most of reptiles, fish, amphibians. Such type of teeth are found on the surface of tongue, in the lining of pharynx and over the gill-arches, obviously to prevent the escape of food through the gill-slits. All teeth of fish are of the same shape. Such a condition is called homodonty and the dentition is called homodont.

Homogenised Condensed Fish A liquid product from whole fish or offal containing about 50% of moisture prepared as an alternative to fish meal.

Homoplastic/Heteroplastic In induced breeding the pituitary extract of a mature fish is injected to spawners. If the pituitary extract of the same species is used it is called homoplastic and if the pituitary gland of a different species is used it is called heteroplastic.

Hook Bent sharpened pieces of steel wire usually with barb for catching fish.

Horizontal Frame Line The lines inserted through the lumen of the framed gill net meshes at regular intervals to make horizontal sections in a frame net.

Horizontal Transmission Infection transmitted between different individuals of a stock or populations.

Horizontal Water Sampler It is an instrument to sample water from desired depth. The cylinder lies in the water with its transverse section completely open, allowing the water with its sediment and still to flow through it undisturbed. The cylinder is held parallel to the current by a stabilizing rudder. The ski-type base is adjustable enabling samples to be taken on the seabed. A messenger closes the sampler when the desired level has been reached and the water temperature is measured simultaneously by a reversing thermometer. The water sample can subsequently be in the laboratory to determine the weight, particle size of the sediment mineral components etc.

Hot Marinated Fish Fish that have been marinated in hot vinegar.

Hot Smoked Fish Fish cured in hot smoke at temperature upto 250°F (about 120°C) so that the protein is coagulated and the product can be eaten without further cooking. Temperature in the fish must reach at least 140°F, various products are Buckling, Kieler, Sproten, Bloater, Smokie.

Howser Laid Rope A rope in which three strands are twisted together without a central core.

Humus The non living finely divided organic matter in soil derived from, the decomposition of animal and plant substances by soil bacteria.

Hybrid A cross bred product coming from parents which diverge noticeably in their hereditary characteristics. Generally hybridization is possible only between strains or sub species, some time also between species which are very closely related to each other or which are not yet firmly established with regard to genetic history. Frequently hybrids are unable to reproduce.

Hybrid Vigour The mating of genetically dissimilar individuals of totally unrelated strains which may yield off spring which are better adopted to survive than either parent strain.

Hybridisation Crossing or fertilization of one species by the sperm from another with the intention of producing progeny with the best characteristics of both parents.

Hydraulic Load The daily rate of flow of water through a given volume or over a given surface area of a filter usually expressed as m^3 of water or m^2 of filter respectively.

Hydraulic Loss It is loss due to pipe wall friction, elbow design, joints, gatevalve, sudden reduction or enlargement of pipe size. This is expressed in its equivalent height or head or water loss.

Hydrilla It is a submerged aquatic plant rooted to the bottom. Stem cylindrical, branched and solid, roots hairy, leaves sessile and in groups.

Hydrodynamic Survey Gathering of information about tides, currents, direction and volume of flow of body of water.

Hydrogen Sulphide (H_2S) A colourless very poisonous gas with on odour of bad eggs.

Hydrologic Cycle The path by which water moves in nature from one physical state to another. Water vapour in the atmosphere condenses and appears as dew or precipitates as rain snow or hail. Liquid water and ice or snow evaporates. The evaporation rate is greater if the water is fresh than if it is salty because the dissolved salt slows down the rate of evaporation. The movement of air in the form of wind, increases it and humidity decreases it. The movements of water from place to place or from one state to another are physical changes. At no time in this process the water molecule H_2O changes.

Hydropic Pertaining to or affected with dropsy

Hyper aemia The increase in blood flow during acute inflammation.

Hyper osmotic The increase in concentration of osmotically active components in a solution.

Hyper plastic Pertaining to abnormal multiplication or increase in the number of normal cells in normal arrangement in tissue.

Hyperbolic Bottomed Tank Deep approximately oval larviculure tank whose bottom has a hyperbolic shape overlong and short axes to aid even dispersal of larvae and suspended food.

Hyperplasia An increase in the number of cells in a tissue due to increased cell division. This occurs in normal growth as well as in response to wounding or tumour growth.

Hypertonic/Hyposomotic A solution or cell which has a lower ionic or solute concentrations. This will give up water to a stronger solution.

Hypobiotic Pathological processes which have in common the decrease in cell metabolism resulting in deficient growth or abnormal patterns of growth.

Hypochromic A condition in which the amount of haemoglobin in red blood cells is lower than normal.

Hypoesthesia Abnormal decreased sensitivity to stimulation.

Hypolimnion Deeper, denser waters of lakes and the ocean during periods of stratification. The volume of cold water that occurs below the thermocline in standing bodies of water.

Hypophysation Hormone injected to induce breeding of fish.

Hypothalmus Part of brain which controls many internal body functions and the activity of the pituitary gland, produces releasing hormones.

Hypotonic/Hyperosmotic A solution or cell which has a higher ionic or solute concentration, this will attract or pull in water from a weaker solution.

Hypoxia The reduction of oxygen supply to tissues below physiological levels .

I

Ichthyo geography The study of fish distribution which results from the ecological relationships of fishes is Zoogeography or more specifically ichthyo-geography.

Ichthyology It is defined as the study of pure and applied aspects of fishes (in Greek, ikhthus = fish; logus = account). Aristotle (384 - 327 B.C.) is said to be the founder of ichthyology.

Ichthyophythiriasis It is caused by ciliated protozoans *Ichthyophthirius multifilis*. Adult stages of the parasites are spherical. At maturity the parasites leave its host and becomes a free swimming form. Adult parasites swim about for 2 to 6 hours and then attach to any suitable substratum and secretes a thin membrane over their bodies which ultimately form the cyst. The adult protozoans undergoes multiple fission to form young individuals called tomites within the cyst. Under suitable condition the tomites will break from the cyst and swim about seeking a fish host. The presence of small white pustules on skin and fins are the characteristics of the disease. Infected fish will congregate near the source of incoming water or in the vegetation.

Imbibition The phenomenon in which a substance absorbs a liquid and swells but does not necessarily dissolve in the liquid. The process is reversible, the substance contracting or drying.

Immunity The ability of plants and animals to withstand harmful infective agents and toxins .

Immunization The protection of an organisms against by exposing the animal to parasite antigens so that its immune system recognizes and produce a rapid response against the same antigens on subsequent exposures.

Immunodepression The decrease in response of the immune system to antigenic materials because of an established infection or exposure to some chemical agents.

Impermeable Having a texture that does not allow water to move through perceptibly.

Impervious A term applied to a material through which water can not pass or passes with great difficulty.

In Vitro Literally "In glass" describing experiments or techniques performed in laboratory apparatus, rather then in the living organisms.

In Vivo Literally 'In Life' describing processes that occur within the living organisms.

Inbreeding Breeding between closely related individuals. The most extreme form of inbreeding is self fertilization, which occurs in some plants. In fishes mating between siblings or between parents and offsprings is generally the closest form of inbreeding. Inbreeding increases the homozygosity so that deleterious recessive genes are expressed more often in the phenotype and decreases hetrozygosity and hence the potential genetic variability of the population. There is also a general lowering of vigour in inbreed stock (in breeding depression) which is especially pronounced amongst normally outbreeding population. In human societies there are usually cultural restraints on marriages between close relatives.

Inbreeding Depression A loss of vigour associated with self crosses or crosses between closely related individuals.

Income Statement A statement showing the firm revenue and the division of its revenue into the costs, taxes, dividends and retained earnings.

Incubation The time elapsing from infection to manifestation of a disease. The incubation time may last only some hours but also some weeks. This depends upon the species of the germ and on the resistance of the infected organisms.

Indemnity The amount payable by the insurer to the insured either in the form of cash; repair replacement or reinvestment in the event of an insured loss which amount is measured by the extent of the insured's pecuniary loss is termed the indemnity. It is set at a figure equal to but not more than the actual value of the subject matter insured just before the loss, subject to the adequacy of the sum insured.

Independent Variables A variable which causes change in another variable. In an equation relating expenditures and income, income is the independent variable influencing the level of expenditures.

Indian Cure Salmon (USA) Brined salmon sides or strips of meat hard smoked for about two weeks at temperature not higher than 21 to 36°C, also called Beleke, Hard smoked salmon, Indian style salmon.

Indian Fisheries Association It was established in 1964. The main office is at CIFE, Bombay. The objective of the association are to promote scientific character of fisheries and its related subjects and to bring together those interested in these pursuit.

Indicator A substance used to test for acidity or alkalinity of a solution by a colour change e.g. Phenolphthalein, litmus paper.

Indicator Species An organism that can be used to measure the environmental conditions that exist in a locality. Lichen-Species are indicators of levels of pollutions as different species are sensitive to different levels and types of pollutants. Tubifex worms indicate low levels of oxygen and stagnant water.

Indigenous Occurring naturally in a country.

Induced Demand The portion of aggregate demand which varies with income, Induced changes in demand are caused by changes in income.

Industrial Fish Fish caught specifically for reduction into meal and oil. In some countries it might also refer to fish used for other processing.

Industrial Water Sampler The industrial water sampler is same as standard water sampler. It is developed to meet the most exact analytical tests, the sampled water no longer comes into contact with metal, thus tests for determination of trace elements are possible.

Infauna Benthic organisms which are partially or fully buried in the sediment.

Infection The diseased condition arising when pathogenic microorganism enter the body, establish themselves and multiply.

Infectious Hypodermal and Hematopoetic Necrosis Virus Disease (IHHNV) This disease affects all stages of *P. monodon* caused by member of family Picornavividae. The affected shrimp shows erratic swimming behaviour rising slowly to the water surface hanging, rolling over untill the ventral side is up. Shrimps would eventually right themselves up, become weak and loose their appetite rise to surface and die.

Infectious Pancreatic Necrosis (IPN) It is a virul infection of trout. It is stable virus, the smallest know fish pathogens of about 20 micron in diameter. The first sign is a sudden increase in mortality. The largest and best appearing fry/ fingerlings are affected first. Whirling is typical behaviour when the mortality is high swimming victims rotate about their long axis, and death usually ensues within an hour or two, symptoms include overall darkening protruding eyes, abdominal distention and at time hemorrhages in ventral areas including bases of fins. Liver and spleen pale. The digestive tract without food and stomach appears whitish. Histologically there is pronounced pancreatic necrosis.

Infiltration The downward entry of water into the soil.

Infiltration Capacity The rate at which water penetrates the soil governed by the texture of the soil vegetation cover and the slope of the ground.

Inflammation A defensive reaction of animal tissues to injury infection or irritation, characterized by redness swelling heat and pain. Histamines released by damaged cells dilate capillaries and causes them to leak plasma and leucocytes. Fibrin forms a clot behind the bite of injury sealing it off and leucocytes attack the infective microbes.

Inflammation of the Swim Bladder An inflammation of the swim bladder caused by bacteria which often appears as a secondary symptom of infectious dropsy. In these cases the swim bladder is filled with an evil smelling liquid, the belly is swollen and the manner of swimming becomes sluggish.

Inflation An increase in the general level of prices, inflation has many causes; cost push inflation results from increases in inputs costs. Demand pull inflation results from an expansion of demand. Expectational inflation results from the fact that when people expect inflation, their behaviour (such as negotiating a wage increase or a long term contract) is altered in ways which will cause the inflation rate to increase.

Ink Blackish coloured liquid released by cephalopoda into the surrounding water when danger threatens. Sometimes used in the sauce when canning cephalopoda.

Inland Fisheries Society of India It was established in 1969. The permanent address of society is CICFRI, Barrackpore, West Bengal. The society is dedicated to the advancement of knowledge of fisheries science and allied fields. It publishes original research works, monographs, invited review articles and books on specialised subjects for exchange and dissemination of information. It also offers opportunities for infusing new interest in fisheries research and in its development activities by way of organising scientific meetings/symposia seminars etc.

Inlet Passage by which water is allowed to flow in

Innate immunity Various host defences against a parasite which are nor dependent on prior exposure to the parasite not they dependent on immunological memory.

Inorganic Without life, of mineral origin.

Insecticide An insect killer made of plants or chemicals of organic or inorganic origin.

Instar The interval in the life cycle of an arthropod between any two moults.

Insurance It is the financial mechanism which aims at reducing the uncertainty of loss by pooling a large number of uncertainties, so that the burden of loss is distributed, Generally each policy holder pays a contribution to a fund in the form of a premium assessed by the insurer commensurate with the risk he introduces which is established and administered by the insurer and out of the fund are paid the losses suffered by any of the insurer.

Insured Peril The cause of the loss which entitles the insured to recover under the policy.

Integrated Integrated applies to a crustaceans farm which maintains or has control over its own support facilities, brood stock and feed supplies, hatchery, nursery, processing, marketing.

Integrated Farming System An integrated farming is one in which waste materials from open enterprise is used to improve production on another. Thus increasing the efficiency of both; e.g. fish-cum-pig, fish-cum-duck, fish-cum-poultry farming, fish-cum-paddy cultivation, etc.

Integrated Fisheries Project The integrated fisheries project (erstwhile Indo-Norwegian project) came into being in 1952 as a result of the agreement between Govt. of India, UNO and Govt. of Norway. The project started to increase the return from fishing, improved fish handling, processing, distribution methods, health and sanitary condition and general improveing the standard of living of the fisherman in the project area. The project expands its activity covering exploratory and experimental fishing, research training of personnel and establishment of integrated fishery complex at Cochin and other places. The project also assisted the state govt. with expertise, equipment and machineries to establish boat building yards processing and fish meal plants and berthing facilities. With the termination of the agreement with the Govt. of Norway in 1972 the Indo-Norwegian project at Cochin become the integrated fisheries under the direct control of Govt. of India.

Intensive Culture High intensity Aquaculture yielding far in excess of natural productivity levels usually involves high stocking rate in purpose built rearing facilities with use of controlled feeding and management of oxygen levels in the water.

Inter Moult The period between each ecdysis during which the crustacean is hard shelled.

Inter Sex An abnormal organism with physical characteristics that are the intermediate between the male and female of the species. Intersex may or may not have both male and female reproductive organs and are usually sterile. They are usually the result changes or malfunctions of the sex chromosomes and/or sex hormones during development.

Interaperitoneal Within the peritoneal cavity.

Interaspecific Among individuals of a species.

Intercellular Situated or occurring between cells in a structure.

Intertidal Zone on a beach, the area between high tide and low tide.

Interface A surface forming a boundary or demarcation between the different bodies substances or phases. For example there are interfaces between water layers of different saline concentrations.

Intermediate Goods A goods which is produced by one firm and sold to another firm as an input.

Intermediate host A host in which there is development of the asexual or immature stages of a parasite.

Internal Environment The medium surrounding the body cells of multicellular animals i.e. intercellular fluid. In vertebrates its composition is kept relatively constant by the mechanisms homeostasis.

Internal Rate of Return (IRR) The interest rate that will just equate the capitalized value of an investments future profits with the investments initial cost.

Interograssive hybridization Cross breeding between two species of fishes that results in extensive back crossing (interbreeding) between the hybrids and individuals from either parent species, and between hybrid themselves.

Interoparity Life history in which a specific individuals may reproduce in more than one reproductive season during its life.

Interspecific Between two different species.

Interstitial Water/Pore Water It is a transition compartment for solutes between soil and water. As pore water surrounds the soil nutrient are released from the soil or are removed from water. The composition and concentration of solutes in interstitial water are the result of concentration gradients and solubility kinetics of the solutes. They encounter together with the movement of the pore water, changes in composition appear with depth. The concentration gradient and mobility of water in soil ensure the transport of nutrient across the sediment water interface. Exchange of nutrients across the interface is the considerable importance for the tropic status and productivity of the pond.

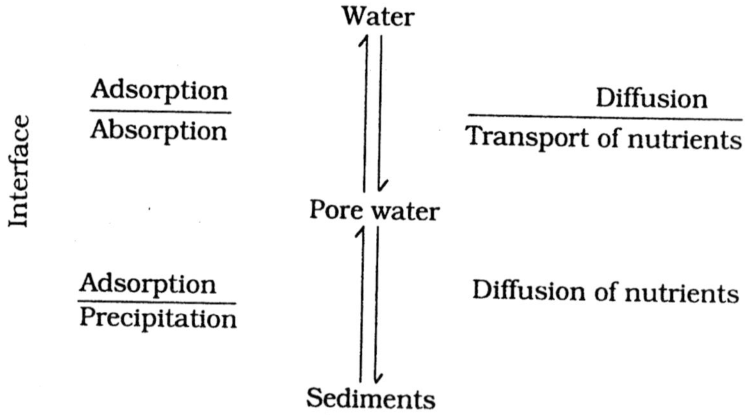

Intramuscular Injection Injection site usually in the dorsal musculature .

Intraperitoneal Injection Injection into the abdominal cavity.

Ipomoea /Water Spinach It is a marginal aquatic plant. The stem is rounded hollow and jointed roots fibrous leaves oblong petiolate arranged alternately, flowers big conical, solitary, Purple white yellow in colour.

IQF Individually quick frozen a processing step in the preparation of high quality crustaceans.

Iron Bacteria Filamentous bacteria. They grow by oxidizing ferrous compounds to ferric oxide e.g. leptothrix.

Irradiation A method for preserving fish by exposure to ionising radiation from radioactive isotopes or an electron source. At pasteurisation doses of 150000 to 450000 rads, over 90% of the spoilage bacteria are killed and the shelf life of the fish at 0 to 20°C is extended by about two weeks.

Isinglass Isinglass is obtained from the inner silvery layer of the air bladder of certain fishes. This layer is separated from the outer thick and fibrous layer of the bladder and treated to produce the isinglass. It is very similar to gelatin because it swells but does not dissolve in water. It is used in confectionary as a substitute of gelatin, in the preparation of adhesives and plasters and as an agent in beer and vinegar industries.

Isobar Refers to equal pressure.

Isohaline Refers to equal salinities.

Isotherm Refers to equal temperatures.

Isozymes Different molecular forms of an enzyme with different physical properties.

Ivory Marine sources of ivory are the toothed whales and the walrus.

J

Japanese Canned Fish Pudding Fish flesh ground and seasoned with salt, sugar and mirio (sweet liquor brewed from rice) boiled steamed or boiled.

Jellied Eels Pieces or steaks of small eels precooked in light brine or vinegar and salt solution packed when cool into gelatin solution or aspic in cans or glass jars.

Jetty A man made rock structure usually at a harbor or river mouth or other area of ship passage, jetties are constructed primarily to direct currents and to inhibit the shoaling or sitting up of navigable channels.

Jib Triangular pieces of webbing attached to both sides of upper and lower edges of the mouth portion of the trawl net also called wedge.

Joining of the Net The process of connecting by means of a thread. There are two methods of joining netting namely sewing and seaming : 1. Seaming (lacing) : The fastening by means of a thread at the edge of each panel without involving the formation of a row of half meshes. 2. Sewing (joining upward) the formation using a thread of a row of knotted half meshes or loops loosely placed through the meshes at the edge of the net.

Jumbo The term used for large specimens of several species, but particularly haddock, skate, shrimps.

Jussiaea/Water Purslane It is a marginal aquatic plant, the stem is creeper type, cylindrical, branched, roots thread like, leaves sessile or petiolate, oval, margin smooth, arranged alternately or oppositely, flowers white or yellow.

Juvenile Post metamorphosis stage of a fish's life cycle, starting once all organ, systems are essentially fully formed and ending once the individual reaches sexual maturity.

K

K-cut Cutting nets in size.

Kakinada Nava It is a narrow round bottom boat without a keel. Drift nets, long lines and stake nets are operated from these boats. They are strong enough to land on hard sand beaches even in surf although they navigate generally through shallow backwaters and muddy passages. This is operated in east Godavari district at Kakinada.

Kedgree Dish made from boiled ricefish or meat, egg cream or butter herbs etc., also canned.

Kelp Brown algae Laminaria spp. harvested and dried as a source of alginic acid also for manufacture into meal for animal feeding stuffs.

Kelt Salmon that has spawned.

Kickling Winding an old rope around a heavy rope to protect it from chaffing.

Kidney One pair of major excretory organs of vertebrates which may also function in osmoregulation. They are made up of nephrons which are responsible for the filtration and selective reabsorption of materials/water, mineral, salts, glucose etc. and the production of wastes.

Kill chill To dip in ice water then precook at 65°C for 15-20 seconds.

Kink A twisted loop in the loop or wire rope. It is generally due to over twisting or winding the cordage against the lay.

Kipper Fat herring split down the back from head to tail lightly brined and cold smoked may be artificially coloured.

Klip Fish Split salted fish from cod species that has been spread on rocks or special platform for sundrying or more usually that has been artificially dried.

Knittles The strand of two ropes twisted together.

Knot (i) A measure of nautical speed. One knot-one nautical miies (6080 feet) per hour. Originally a log attached to a rope with evenly speed knots on it was heaved over board to measure the speed of a ship. By counting the number of knots let out in specific time period, standardised at 28 seconds the sailor could calculate the speed hence the term "Knots" . The space between knots was standardized at 47 feet 3 inches.

(ii) A knot is a tie made by one end at some parts of its own body to make a loop or over some other object to get it fastened to it.

Knot Disease Small knots the skin of common carp. It is caused by protozoa *myxobolus exigus*.

Knot Slippage Sliding or slipping of the knots of the webbing caused by bad workmanship or by the smoothness of the material with which the webbing is made.

Knot Stability It is the ability of the knots in the netting to retain their original form by resisting the inversion into another form without slip and also the loosening with resulting slip but without inversion.

Knot Strength It is the tenacity of a single fibre or yarn in which a simple knot has been tied.

Knotted Netting Netting made with knots.

Krill It looks like small shrimp but does not have the characteristically abrupt "shrimp bend". They have large black eggs and under the abdomen are rows of light organs which can be lit up or extinguished. They can reach 5 cm. long and live for 2-3 years. Krills are caught with special trawls and fine mesh dip nets allied with lights.

Krupuk Ground shrimp (other fish also) mixed with tapioca flour, salt and seasoning kneaded with water and steamed in moulds, then cooled, cut into slices and sundried, swells and becomes crisp when fried in deep fat.

Kype Hooked jaw, a secondary sexual characteristic exhibited by many mature male salmonids.

L

Labour Productivity It mainly refers to the products or production value created by one labour in the period of one year or one productional cycle. The formulae for calculating labour productivity is as follows.

Labour productivity Rs/labour/yr. or Kg/labour /yr.= Total products of fish crop and animals etc. (kg) or total

$$\frac{\text{production value of all activities (Rs.)}}{\text{Labour consumption (manday / yr.)}}$$

Lacustrine Lake ecosystem.

Lamayo Fish product consisting of salted partially dried macerated shrimp. It is common in Philippines.

Lamelliform It is characterized by the presence of a large inter-branchial septum projecting beyond the length of lamellae. This type of gill is known as lamelliform and is found in cartilaginous fishes (lamellibranch).

Laminarin Carbohydrate extract roughly equivalent to the starch in the land plants obtained from brown algae (Laminaria spp.)

Lanceolate Tapering at both ends.

Land Scape The sum total of the characteristics that distinguish a certain kind of area on the earth's surface and give it a distinguishing pattern in contrast to other kinds of areas.

Large Volume Water Sampler (LVWS) It is a sampler designed in the order to collect large water samples necessary for the determination of 14_C, 90_{Sr}, and 85_{Kr}.

Larva The pre adult stage of many organisms which does not structrally resemble the adult. The larvae usually exploit a different food source from that used by the adult.

Larval Mycosis in Prawn It is caused by the fungus *Lagenidium* sp. generally in eggs larvae and post larvae of *P. monodon*. The infected eggs and larvae appear whitish become weak and may eventually die. These aquatic fungi produce highly motile zoospores that can easily invade other host. The fungal hyphae replace the internal tissues of the shrimp and extend outside the shrimp body to form discharge tubes. Infected eggs donot hatch and larvae lose equilibrium and exhibit respiratory difficulties. Heavy mortalities upto 100 % within two days, may occur.

Lastridge The line along which the top and bottom parts of a trawl are locked together. Also called lace hoods.

Lateral line It is a line formed by a row of scales having pores (holes) running horizontally along the sides of the body of a fish from behind the gill-covers upto the tail. It is a sense organ by which the fish perceives vibrations.

Laver Bread Food stuff prepared from red algae (Porphyra spp.) by boiling in the salt and cooling, mincing and dyeing, often fried before eating.

Lay The direction of strand or helix;

(i) Cross lay : A multiple layer strand in which the wires in the covering layer repeatedly cross over in the inner wires.

(ii) Lang's lay : The rope in which the lay of the wires in the strand is in the same direction as the lay of the strand in rope.

(iii) Regular lay : A rope in which the lay of the wires in the strand is opposite in direction to the lay of the strand in the rope.

Laying Refers to twisting of twine cable etc.

Laying Out Putting a net into the water from fishing purposes.

Lazy Line Rope attached between the head line of a trawl and cod end, the cod end used to facilitate drawings the cod end towards the vessel.

LD 50 Median lethal dose i.e. the dose of toxin at which 50% of exposed animals are killed. It is used as a standard measure of toxicity.

Leaching It is the process by which water percolates down wards due to gravitational movement of water in the soil. By this water movement much of the portion of plant nutrients are drained with water. It is measured by the following flow of water :

1. Very small		0.127 cm/hour.
2. Small	—	0.127 – 0.8255 cm/hour
3. On average small	—	0.635 – 2.032 cm/ hour
4. On average much	—	2.032 – 6.35 cm/hour
5. much		6.35 – 12.7 cm/ hour
6. very high	—	More than 12.7 cm/ hour.

Lean Fish Fish containing 2% or less of fat or oil in the flesh in live condition.

Leather carp A variety of common carp *cyprinus carpio* characterized by almost complete absence of scales.

Lemna It is a floating aquatic plant, stem modified to flat rounded 2-12mm in size, purple beneath, root fibres 2 or more, leaves absent, flowers small. It has one root to each joint and is green beneath.

Lemniseus One of a pair of tegumental projections into the anterior portion of the acanthecephalan, pseudocoelom.

Length of Lay The length (in mm) of a complete turn of one and the same strand measured in a straight line parallel to the axis of a rope.

Leptocephalus Flat transparent ribbon shaped young one of an eel.

Leptocephalus Larva The protecephaline larva hatches out from the egg of Angueilla (eel). It bears a yolk sac. As the yolk sac is gradually absorbed the protocephaline larva changes into a Leptocephalus larva. The body is glassy transparent, flattened, leaf-like and vary in length from

5 cm to 180 cm. The larva swims with eel like undulatory movement of the body and exhibits diurnal vertical migration, approaching the surface only at night. It takes nearly two and half years for leptocephali to migrate passively from their spawning grounds in sargasso sea to the nursery grounds in the coastal waters of Europe.

Leptoid scales In majority of bony fishes, scales are of the ganoid origin but have lost the ganoin layer completely and there are no pulp-cavities. Thus, the scales of these fishes are left only with the thin isopedin layer and are known as Leptoid scales. There are two kinds of leptoid scales, i) Cycloid and ii) Ctenoid. These scales differ from each other in a relatively minor way, involving the presence or absence of ctenid on the free posterior edge. Leptoid scales with Ctenid are called Ctenoid. In both, the anterior part is overlapped by the posterior edge of another scale. Exposed part of cycloid scale has chromatophores. In both types there are distinct lines of growth. Fish scales are never shed.

Lernaeasis/Anchor worm The female parasite is needle shaped. The body is 6-12.4 mm in length and consists of head, thorax and abdomen. There is no distinctive demarkation between the three sections. On the head there is a pair of dorsal horns and a pair of abdominal horns. They function as anchor. The male is smaller and occasionally parasitic. Initially the diseased fish behaves uneasily has a poor appetite is thin and moves slowly. The areas that parasite has penetrated are inflammed and swollen and tissues are necrotic. The wounds are often invaded by saprolegnia. On a juvenile fish four or five Lernaea 6-9 cm long can be lethal. On a young fish just one or two parasites can retard growth.

Lesions Abnormal changes in tissues of body functions.

Lethal Deadly

Life Cycle The sequence of changes making up the span of an organism's life from the fertilization of gametes to the same stage in the subsequent generation.

Lift Net Net fixed to a circular, square or cross shaped frame which is positioned above it. Sometimes attached to a wooden pole or fixed frame to assist in lifting and some times baited.

Lifting Line The line used in purse seine operations for lifting a group of floats of limited number, while in operation if required and during hauling.

Light Cure Fish treated with only small quantities of salt (e.g. 16 to 20 parts per 100 of fish) or left in salt for a short time only (e.g. 3 to 5 days in pickle) product contains 20-30% salt on a dry basis, cod and other white fish salted this way.

Light Fishing A method of fishing involving use of lights to attract the fish.

Lime See calcium oxide, calcium hydroxide.

Liming of the Soil It not only corrects acidity but also improves response to fertilizers, promotes desirable bacterial activity, improves soil structure and facilitates the growth of green manure.

Limited entry Regulations that limits the number of participants in a fishery. It is usually applied in a manner that identifies specific participants.

Limnetic zone In offshore regions of lakes too deep to support rooted aquatic vegetation the portion of the water column under the surface that has sufficient light to support phytoplankton production.

Limnology The scientific study of freshwater and its flora and fauna. It also includes the studying of the chemical physical aspects of inland waters.

Linpe Method The method is named on scientist R.E. Peter of Canada and H.R. Lin of China, who proved that dopamine interferes in synthesis of gonadotropins. A combination of LH and dopamine antagonist can be effectively used as a substitute to fish pituitary for carp breeding.

Liquefaction The result of necrobiosis conversion of tissue into a semisolid or fluid mass.

Liquid Manure Animal waste from intensive animal husbandry units, where no bedding material is used. It is usually collected as a slurry.

Litmus Paper Red or blue acid alkali indicator paper. In acids blue litmus paper turns Red. The paper is prepared by soaking absorbent paper in litmus solution and then drying it. Litmus solution has a pH range from 5-8.3. It is obtained from a Lichen (*Lecanora tartarea*).

Littoral The coastal zone

Littoral Zone Shore line region on lakes where light intensity at the submerged substrate is sufficient to allow growth of rooted aquatic vegetation.

Live fish A fish that is capable of living outside water for some time by virtue of its capacity to breath atmospheric air.

Loam The textural class name for soil having a moderate amount of sand, silt and clay. Loam soils contain 7 to 27% of clay 28 to 50% of silt and less than 52% sand.

Locks Mild cured sides especially of king salmon, cold smoked for one to three days, also frequently called lox, common in USA.

Long seasonal pond A water bodies that is available for fish cultivation during six to eight months in a year.

Lophorbranchs In sea-horse and pipe-fishes, the gill filaments are replaced by curious tufted processes, such gills are called lophobranchs.

Loss Adjustment Following a notification of a claim the insurer must determine the extent to which loss has occurred and the amount which can be paid to the insured as indeminity. This process of gathering information and making a decision is called loss adjustment or loss assessment.

Loss Frequency The rate of the occurrence of losses often expressed in terms of the number of incidence over a period of time. It is one method used for measuring the effectiveness of loss prevention services.

Lugol's iodine This is recommended for preservation of planktons, the cilia and flagella of plankton absorb iodine and protozoans sinks to the bottom of the container. Lugol's iodine solution can be prepared by dissolving 200 grams neutral potassium iodine of analytical quality and

1 gram iodine in 200 ml of distilled water. When the iodine is completely dissolved add 20 ml of glacial acetic acid.

Luminous Vibriosis A bacterial disease which affects eggs. larvae and post larvae of *Penaeus* sp. caused by *Vibrio arveyi, V. splendidus*. This is easily recognised by the characteristic blue greenish luminescent appearance when observed in total darkness. Larvae become weak and opaque-white resulting mortalities.

Lures It is an artificial bait to attract the predatory fishes to the hooks in such a manner as to stimulate a fish in motion while gear is in operation.

Lute fish/ Alkaline Cured Fish Product prepared by soaking stock fish for several days in solution of soda and lime and then for several days in water to remove chemicals.

Luteinizing Hormone (LH) A glycoprotein hormone secreted by the anterior pituitary lobe under regulation of the hypothalamus. In female it stimulates ovulation and in male stimulates interstitial cells in the testes to secretes androgens.

Lymphocystis Disease This is a viral infection, the external lesions of lymphocystis are raised growths of generally granular, wart like or nodular tissue which is composed of several to many greatly enlarged host cells and their covering membrane, colour of the lesion is usually light and may be white grey or creamed colour. Larger lesions may show pink to vascular net work and matured lesion may become slightly hemorrhagic.

Lyre A grade of isinglass.

Lysin An antibody which causes cell lysis or a bacterial toxin which lyses cells.

M

Mackrel Style Split Fish Fish that have been split along the back leaving in the back bone.

Macro-vegetation Larged sized plants as opposed to microvegetation or minute plants.

Macrobenthus The larger organisms (from 1 mm upward) living in or the pond bottom.

Macrophagous Applied to animals which feed at intervals on pieces of food which are in relation to their body size.

Macrophytes Large aquatic plants rooted or floating.

Macrospores Large spores in microsporidia characterized by spore di or polymorphism.

Macrothrix Belongs to Phylum Arthropoda order cladocera, family Macrothricidae. The head is large, its dorsal margin rounded overabruptly into anterior margin 1st antennule long and freely moving 6 pairs of feet, abdominal setae present.

Main Body The portion which forms the bulk of the purse seine net and lies between the wing and the bunt.

Main Line /Long line The rope to which the branch lines or snood with the hooks are attached.

Mainstream The main channel of a river basin into which tributaries empty.

Major carps One of the larger carps attaining a length of over two feet.

Makassar Fish Anchovies headed, salted for several days and then mixed with rice, yeast and spices, packed in bottles in the resultant red pickle.

Malachite green A green mineral consisting of copper carbonate and hydroxide used to remove fungi and some protozoans from fish.

Mamrout Fermented fish paste made from flesh and entrails, other kinds of fermented fish paste or mam are also made in Vietnam.

Management Hatchery Planning and supervision of hatchery and farm operations, assessment and manipulation of fertilization, feeding and water exchange to control plankton density within desired limits.

Manca The larval stage of a parasitic isopod it has six pairs of legs and usually has good swimming ability.

Mangroove A tidal salty marsh community dominated by trees and shrubs mainly Rhizophora spp. It cleared for pond construction the underlying soil is usually found to be strongly acidic.

Manila Hard fibre taken from the leaf stem of Abaca plant, musa textiles grown abundantly in the Philippine islands, strong twine, cord, ropes etc. are made from abaca fibre.

Manure Coefficient It can be calculated as,

$$\frac{\text{Total net increment of fish body weight}}{\text{Total consumption of food / manure}} \times 100$$

Manure Conversion Rate It can be calculated as

$$\frac{\text{Consumption rate of manure}}{\text{Net increment of fish body weight}}$$

Marginal Cost The change in output resulting from a one unit increase in the quantity of the variable input.

Marginal Revenue The change in revenue resulting from one unit change in output.

Marginal Utility The change in total utility or satisfaction resulting from a one unit change in the consumption of the good.

Mariculture Farming of organisms in the sea.

Marinade Fish especially herring cured in acidified brine with or without spices in special containers or barrels after curing, packed in mild acidified brine also with spices, vegetables or other flavouring agents e.g. Bismarck herring.

Marine Biological Association of India Establised in 1958. The permanent address is Shanmugham Rd. Ernakulam Cochin. The objective of the association is to create active interest in the field of marine sciences and allied, organising lectures, symposia, seminar etc.

Marine Biology Study of life in the sea; also called biological oceanography.

Marine Oils Fish oils characterized by a high fatty acid content. The fatty acid that constitute these oils are carbon compound whose molecules consist of chains of carbon atoms to which hydrogen and oxygen atoms are attached by chemical bonds. The natural marine oils have a variety of uses. They are commercial interest as edible oils as drying agents for the plant and coating industry and in the tanning of leather.

Marker Buoy A buoy attached to a part of gear to mark its position at the surface.

Marker Floats Floats specially used to fix the gillnet at desired depth.

Market Related Risks Risks due to loss of consumer appeal, health regulations actions of third party (the marketing middleman).

Marl A general term for calcareous clay or calcareous loam.

Marling A soft laid two ply cord generally used for binding fibre rope ends.

Marling Hitch A series of round turns in which the end is passed over the standing part and under the bight and pulled taut in each turn. Unless there is an eye in the end of the rope, a marling hitch is usually started with a timber hitch. This is used for long lasting sails etc.

Marsh Broad wet lowland at least periodically inundated with water with substantial stands of herbaceous terrestrial vegetation such as grasses and bushes.

Marsilia/Water Clover It is a marginal aquatic plant, stem is creeping or floating roots minute hair like leaflets 3-5 with entire margins, covering on a long petiole, flowers primitive sporocarp.

Marsupium The chamber in which eggs are brooded in an isopod and it is formed by Ostegits.

Masculinisation Modification of gender of fish by hormonal or genetic means.

Master Float Float attached to the distal end of the fleet end rope to enable location of gillnet fleet end.

Master Sinker A heavy weight used at the gill net fleet end so as to enable controlled drifting in a drift net and setting in set nets.

Masula Boat It is a large open boat used along the coromandel coast and other parts of the east coast. It is used for operating shore seines, steering is done with a powerful paddle sweep 15 feet long. The planks are joined together and stiched with coir yarn. They have frames and floors.

Matsyafed The Kerala state co-operative federation for fisheries development limited is the apex body of fishermen's co-operative. The immediate objectives of Matysafed are

(i) Cooperatirisation of traditional fisherman

(ii) Increased fish production in the traditional sector.

(iii) Financial return to fisherman through beach level auctions.

(iv) Imparting intermediate technology to the Traditional fisherman.

(v) Development of infrastructural facilities.

Maturation Ripening the cell divisions by which gametes are produced to enter a phase of reproduction competence.

Mean Depth It is defined as the volume of water bodies divided by the area. It is considered the most important morphometric parameter.

Mechano Receptor (Organ) Organs in fish (mainly along lateral line) which enables the fish to sense currents and its own movement.

Medium Salted Fish Fish cured by using 20-28 parts of salt to 100 parts of fish by weight, product contains 30% to 40% salt on a dry weight basis.

Mehlis gland A gland which evacutes into the otype in the digenean female system.

Melanophores Dermal cells containing melanin

Mending the Net Repairing the holes or other damage in the net.

Merogony The sequence of asexual multiplication (syn. schizogony) to produce a great number of vegetative parasite stages.

Meront An asexual developmental stages to produce merozoites.

Mesh Interspaces of a fixed dimension formed by a sequence of loops or bight (half meshes) which are interwoven by knots.

Mesotrophic Aquatic habitat with intermediate organic load.

Metabolic Disease of Fishes Diseases of the metabolic system with non-parasitic causes. Diseases of the digestive system set in especially after long lasting one sided feeding or sudden changes in the manner of feeding. Therefore those are avoided by variety in feeding in accordance with the preferences of the respective species. The most common result of wrong feeding method is the overfattening of the intestines. Cures are generally much more difficult.

Metabolite The product of any chemical change occurring in living organisms, waste metabolites from one species may be used as food by another.

Metacercariae Encysted cercariae of digeneans.

Metamorphosis The marked change in form that occurs between life cycle phases during the development of crustaceans.

Metaplasis A change in tissue type.

Metatrypanosome The trypomastigote stage of a trypanosome that is formed in the vector after cyclopropagative development.

Meter Wheel Meter wheel is used for counting the length of wire paid out from a sounding machine.

Methaemoglobinaemia Under conditions of poor water quality the hemoglobin in cat fish blood can combine with ammonium nitrate to cause brown blood disease due to resulting methaenoglobin.

Methyl Orange An acid base indicator that is red in solution below a pH of 3 and yellow above of 4.4. As the transition range is clearly on the acid side, methyl orange is suitable for the titration of an acid with a moderately weak base, such as sodium carbonate.

Methyl Red An acid base indicator that is red in solution below a pH of 4.2 and yellow above a pH of 6.3. It is often used for the same type of titration as methyl orange but the transition range of methyl red is nearer neutral pH 7 than methyl orange. The molecules are structurally similar.

Metric Number It indicates the length of a single yarn in meters per mass of a kilogram.

Micracidia The larval stage that hatches from a digenean egg.

Micro Encapsulated Diet Very small particles containing compounded ingredients and surrounded by a digestible coat.

Microbial Food Loop The recovery of energy from organic matter to the benefit of the food web. The term was coined to designated trophic links between bacteria, heterotrophic flagellates and the microzooplankton in water. This loop should however be considered as a integral part of the food web of any ecosystem and not a source of sink of carbon.

Microcystis It is a phytoplankton belongs to order chroococales family chroococeaceae. Unicellular, cells variable in shape oftenly colonial.

Microcytic A condition in which red blood cells are smaller than normal.

Microgram Unit of weight, 1/1000 of milligram.

Microparticulate Diet Very small particles of food manufactured for larvae.

Micropyle An aperture in egg through which sperm enters for fertilisation.

Microscopic Minute usually visible only under a microscope.

Microsporic A myxoporean species whose plasmodia forms various types of spores.

Microsporida Smallest protozoans parasite.

Microspridiosis in Prawn The affected organs tissues which turns opaque white indicate microsporidia endoparasitic protozoans to be the causative organisms of the disease in the juvenile and adult of *P. monodon, P. indicus*. The infection rate is relatively slow but the spores and other stages of the parasite replacing the affected tissues are highly pathogenic resulting sterility of spawners with white ovaries.

Mid Water Drift The method of operation of gillnets when allowed to drift in any region between the surface and bottom of the water.

Mild Cured Fish Fish that have been salted or smoked lightly and have limited keeping quality.

Mild Smoked Fish Fish that have been smoked cured for only a short period to develop slightly smoky flavour also called light smoked fish.

Milt Common name for sperm or spermatozoa.

Mimicry Resemblance to other organisms which are protected against enemies by means of venom stings or other protective organs which have come forth during the genetical history of development though it the respective animal may enjoy the advantages which the immitated live being has. The adaptations of an animal to lifeless objects by means of colour and shape is called mimesis. Through this the imitating animal may find protection against enemies, but on the other hand if it is a predator it may also escape by visual detection by its prey.

Mimosine The leguminous plant *Ipit ipit* or subabul contains a toxic substance known as mimosine which is a free amino acid present at about 3-5% of the dry matter.

Minosine is considered to affect the production of thyroxine and hence the growth of the organism resembles structurally with the amino acid. Tyrosine and it might function as an antagonist to this amino acid. Minosine content can be reduced by soaking in water for 24 hrs.

Minced Fish Minced fish flesh used for further processing, free from skin and bones.

Minnow A freshwater fish which generally does not attain a size larger than that of a fry or fingerlings of larger carps.

Minor carp One of the smaller carps as opposed to the larged sized major carp.

Mirror carp A common variety of common carp *cyprinus carpio* having large shining mirror like scales on the body.

Mittes Small piece of tuna meat obtained when cooked fish is cut into slices.

Mixed Compost It consists of green manure and animal waste.

Mixed farming The practice of cultivating more than one species of compatible fishes in the same pond at the same time.

Mixed Layer Depth Thickness of the surface layer in the ocean from surface to the top of the thermocline.

Moina Belongs to phylum Arthropoda order Cladocera Family Daphnidae. Body rounded with heavy abdomen, head large, rounded in front with depression above eye, antennules long and freely movable abdominals setae very long.

Moinodaphnia Belongs to phylum Arthropoda order cladocera family-Daphnidae. The head small antennules on ventral surface of head minute spines on ventral margin, sharp angle but no spine at junction of dorsal and ventral margin.

Mollusks Soft, bilaterally symmetrical animals with head (and brain), foot region coelom, coiled visceral mass, and usually an external or internal shell.

Monk A water control structure governing pond depth at a pond exit and water flow rate at a pond entrance. Screens prevents escape of stock and entry of predators and competitors, same function as sluice gate.

Monoculture Culture of a single species of fish.

Monocyst A membranous and/or fibrous structure located beneath the surface membrane of a ciliate or a flagellate. It is discharged through an opening in the pellide as a mucus like mass.

Monodon Baculo Virus (MBV) The disease is caused by Baculovirus, affect all stages of *P. monodon*. The most obvious clinical signs are pale bluish grey to dark blue-black colouration and yellowish white hepatopancrease, sluggish and inactive swimming movement with loss of appetite, retard growth, destruction of hepatopancrease and lining of digestive tract weaken the shrimps and lead to gradual mortalities.

Monofilament Single filament which is strong enough to function alone as a single yarn.

Monosex Culture The rearing of a single sex in an attempt to reduce size variablity among harvested populations, employed in fish culture to avoid uncontrolled reproduction.

Monoxenous A life cycle which involves one host.

Mooring Rope A synthetic or hard fibre rope used for tying the vessels to the moorings.

Mother of Pearl Nacreous layer on the inside of the shell of a number of molluscs, used as an ornamental material.

Mould Superficial growth of fungus, also used for particular types of fungi.

Moulting The act of casting the exoskeleton.

Mounting Twines Twine used for the attachment of the webbing to the rope.

Mouth Breeders Comprehensive name for fishes belonging to different families which have the common characteristic of breeding their fry in their mouths. Cases of mouth breeding are found in numerous catfishes.

MPEDA Marine product export development authority is a statutory agency responsibile for the planned and regulated development of the Indian sea food industry, specially export production and export promotion. MPEDA provides a comprehensive package of services to Indian exporter and overseas exporter ranging from establishing buyer seller contacts right up to ensuring that products are delivered to importers in prime condition and on time.

MS 222 Tricane methanesulphonate is a light anaesthetic are mostly used during fish transport.

MSY Maximum sustainable yield An average over a reasonable length of time of the largest catch which can be taken continuously from a stock under current environmental condition. It should normally be presented with a range of values around its point estimate.

Mucus A slimy substance produced by goblet cells in mucous membranes of animals. It is viscous and insoluble consisting mainly of glycoprotein. Its function is to protect and lubricate the surface on which it is secreted.

Mucilaginous Slimy

Muck Highly decomposed organic material in which the original plant or animal parts are not recognizable. Contains more mineral matter and is usually darker in colour than peat.

Muck Soil Organic deposits in the soil are classified on the basis of the amount of organic matters that they contain. When the quantity of this constituent ranges for 20 to 50 per cent such a soil is called muck.

Mucoid Feeding Feeding by means of mucus which is extruded from the mouth, then ingested with small particles that have stuck to it. A method employed by some molluscs.

Mucous glands These glands cells are unicellular. They are oval or pear shaped with a neck through which they open outside the epithelium. They are present throughout the epithelium i.e., gill arch, gill filament and secondary lamellae. They secrete mucous, which is glycoprotein, both acidic and neutral. They act as protective, friction reducing, antipathogen, help in ionic exchange, gas and water exchange.

Mucous Stimulator Mucous coat covers the body surface is a protective layer. The chemicals which stimulates the mucous secretion is mucous stimulator e.g. salts, potassium permanganate.

Mucron A cuticular spine at the posterior end in a nematode.

Mulletry A fish farm where mullets form the important crop raised.

Multifilament Filament yarn having several number of individual filament.

Multiple of Weight gain : It can be calculated as

$$\frac{\text{Total fish yield of pond}}{\text{stocked quantity}}$$

Multiple Spawning Multiple spawing is the timely harvesting of the mature gametes repeatedly by more than two spawnings. The fishes are bred by adopting routine hypophysation techniques. Major carps have been breed as many as four times between March and September.

Multiple Weight Increment This can be calculated as,

$$\frac{\text{Fish yield of a certain species (Kg.)}}{\text{Stocked quantity of the fish (Kg.)}}$$

Murrel One of the common live fishes with the head remotely resembling that of snake and hence called snake head (Ophicephalus)

Muscle Necrosis in Prawn Opaque white areas on the abdomen blackening on the edges of the uropods followed by erosion liquid filled boils at the tip of uropods in advanced stages occur in post larva juveniles and adults in *P. monodon.* It is due to stressful environmental condition like low oxygen levels, temperature or salinity shock, over crowding and severe gill fouling.

Muscle System of Fishes The muscle of the trunk, used for locomotion, take up the main space in a fish's body. They stretch from the occipital bone to the origin of the tail fin in double rows that are located on the both sides of the body axis. The upper, dorsal, part of the rows of muscles is separated from the lower, ventral, part by horizontal conjunctive tissues. The different parts of the muscles which on their part too are separated by conjunctive tissue, are inserted into one another life pointed paper bags and fixed to the spine. The muscles that move the limbs , that is the fins, by erecting and dropping them, have developed from parts of the muscles of the trunk. The muscle of jaws system and of the opercles form an independent group.

Mysid A small swarming crustacean (e.g. Neomysis integer) commonly used as fish food in home display aquarium but also a good food for large crustacean larvae such as lobster larva.

Mysis The stage between zoea and post larva in the development of penaeid shrimp.

Myxosporidiosis This disease is caused by Myxosporidian parasites of the genera myxobolus, thelohanellus, myxidium, henneguya chloromyxum etc. The parasite attacks fishes belonging to the size range of 41-150 mm. The pathological effects of the disease in fish are weakness, emaciation, raising of the scales along their posterior margins, falling of scales, loss of chromatophores, perforation of the scales.

N

N-Cut Ref cutting to shape.

N-Direction See direction of netting.

Nacre The iridescent inner layer (mother of Pearl) of many mullusc shells.

Najas It is a submerged aquatic plant. The stem is slender and much branched, leaves small, sessile, ribbon like enlarged at bases leaves toothed arranged alternately or whorled.

Nandus nandus The body is slightly elongated, perch like, body is strongly compressed laterally. Even the big head is nearly covered with scales. The mouth opening is big and may extend far to the front. Palate jaws and tongue are covered with numerous small and pointed teeth.

Nape The back or upper part of neck.

Naping (U.K.) Cutting through the nape that is the flesh between the head and the belly of the fish such as cod as a preliminary to gutting so that when the belly is subsequently slit longitudinally to the vent, the belly wall can be laid open to expose the belly cavity for cleaning. Single naping is the severance of only one half of the belly wall, double naping means cutting through both halves.

Narcotic A substance that induces stupor or sleep.

National Institute of Oceanography It conducts research on marine resources and exploitation of the ocean resources. The NIO sponsors research expedition to the Antartic.

Natural Cure Salt cod product in which the salted fish are washed and drained only long enough to remove excess water before drying naturally for several days. Salt content after curing is about 20% (minimum 17%) also called national cure (Portugal).

Natural Productivity The natural productivity is the capacity of the living community to increase its biotic load overtime. The community of living organisms can be separated according to their position in food web as producers, consumers and decomposers. The separation is based on their mode of harnessing . The energy needed for their metabolic activities. Three main processes emerge photosynthesis, the use of solar energy to elaborate organic compounds grazing and predation or consumption of living organisms containing high energy compounds decomposition or consumption of dead organic matter again to make use of high energy compounds. These three processes are complementary and interdependent.

Nauplius The early larval stage of copepods, crustacean and barnacles, it has three pairs of swimming appendages.

NBFGR National Bureau of Fish Genetic Resources conducts research on cataloguing of fish germplasm resources of India genetic characterisation of fish germplasm identification and development of conservation methods for endangered, threatened and rare species that needs to be conserved.

Necrosis Area of dead cells or tissue in an otherwise living or healthy organ.

Necrotic Pertaining to cell death and enzymatic degradation in a limited area in an organism.

Nekton Mobile aquatic animals which have the ability (in spite of currents and waves) to actively determine their course of movement.

Nelumbo It is commonly known as lotus. It is an emergent aquatic plant. Stem creeping, rhizome, root tuberous long deeply rooted, leaves saucer shaped, bluish green shedding water if pushed below the surface. Flowers with many yellow petals, fruits round and nut like in a light spongy top shaded structure.

Neomycin An antibiotic produced by the bacterium Streptomyces fradiae. It is an inhibitor of protein synthesis affecting gram positive organisms. It acts by binding to ribosomes, resulting in a misreading of the genetic code.

Nepa/Water Scorpion Insect of order-Hemiptera Family Nepidae. Blackish dull in colour prey in a position head downward and forelegs heldout in reptorial position, rostrum short and strong, first pair of wings cover the body two long respiratory filament at the tip of the abdomen grow upto 17 mm in length.

Nerve System of Fishes Fishes have an elongated relatively small brain in which especially those parts are developed which serves to operate the sensory functions. The five sections that are typical of the brain of vertebrates are already developed of the twelve nerves that go out from the brain, the following should be mentioned the paired nerve of smell which originates in the olfactory lobes of the fore brain the seeing nerves which comes from the vision hills on the intermediate brain, the acoustic nerve which established a connection between the hearing organs and the sensorial organs of the skin, the nerve of the taste and a series of other functional nerves for operating the muscles the gills, the heart and the intestine. The spinal medulla is enclosed by the spiral column. Its main function is to transmit the stimuli coming from the brain. According to the number of vertebrae the paired nerve branch out from it forming the unspontaneous sympathetic nerve.

Net Out Put Value of the Per Unit Product This reflects the economic benefit of a farming system. It can be calculated as,

$$\frac{\text{Net output value (net income)}}{\text{Total sum of the cost}} \times 100$$

Net Output Value/Net Income Per Unit Area This reflects the economic benefit of any farming system, it can be calculated as,

$$\frac{\text{Net output value of net income}}{\text{Cultured or cultivated area (ha.)}}$$

Net Protein Utilisation The measure of the ability of an animal to digest and assimilate protein for growth.

Net Rope/Gill Rope Float line made up of two ropes of opposite twist to prevent kinking. The floats are often roved on to one of them.

Netting Yarn It is a term used for material which is suitable for manufacture of netting of fishing nets and which may directly knitted into netting by machine or by hand without having to undergo further process.

Neuromast system The neuromast system is a set of senseorgans of fishes. This system consists of a series of tiny cups lying in linear arrangement in the skin of the head (cephalic canal system) and along the entire trunk region to the tip of the tail (lateral line system). Each cup or neuromast is made up of epithelial cells surrounded by sensory nerve endings. Vibrations in the water stimulate the neuromast or lateral line system and information is utilized like sound. It also serve as depth receptors or the rheoreceptors. In most cartilaginous and bony fishes the neuromasts sink within the skin into depressions, grooves or canals. The canals contain mucous and open on the surface by many minute pores. The cephalic canals are supplied by branches of VII cranial nerve, while the lateral line system by the very long ramus lateralis of the X— cranial nerve.

Neutralization The process by which an antibody neutralizes the infectivity of a parasite.

New Water Any water which is free of organic waste and comes from springs, wells, water mains or newly fallen rain water. While old water which is overcharged with waste product acts restrainingly on growth and metabolism, new water stimulates these.

Nitrate One of the inorganic fertilizers.

Nitrification The aerobic bacterial oxidation of toxic nitrogenous metabolites, ammonia and nitrate to much less toxic nitrate occurs in biological filters also nitrifying bacteria.

Nitrifying organisms Microscopic organisms that convert ammonia and nitrite nitrogen to nitrate nitrogen, a form of nitrogen highly utilized for the production of protein molecules by green plants.

Nitrofuran Synthetic chemicals/drug with antimicrobial activity used for the treatment of some infectious diseases (e.g. Sulphanamide).

Nitrogen A gas forming nearly four fifth of the atmosphere.

Nitrogen Load The amount of nitrogenous waste (usually ammonia) that has to be oxidized to nitrate by a biological filter in a given time in order to maintain acceptable levels in the culture water.

Nobbing Removing the head and gut from fatty fish such as herring by partially severing the head and pulling the head away, together with the attached gut the roe or milt is left in e.g. cut herring.

Non-Adhesive Eggs Eggs which do not attach to any substratum or each other. It can be divided into following types depending upon their specific weight:

(i) buoyant eggs (Specific weight less than water)

(ii) floating eggs (specific weight slightly higher than water)

(iii) semifloating eggs (Specific weight higher than water)

(iv) rolling eggs (specific weight higher than water)

Non-predatory One (fish) that does not kill others for food.

Normal Return Corresponds to an economic profit of zero. It represents the rate of return equal to the return the firm could make in the most profitable alternative use of resources.

Norsels/Ossels Short piece of twine or line about a foot long used to attach the hanging line to the head line at regular intervals of meshes.

Norwegian Cured Herring Hard cured herring made from fresh fat Summer herring that have been kept alive in sea to empty stomaches, sometimes packed in bulk for further processing, special quality.

Norwegian Silver Herring Light cured herring product differ from matje cured herring in that it keeps well at normal temperature no longer produced in Norway.

Nostoc It is a phytoplankton belongs to order Pleurocapsales family-Nostocaceae. The genus is colonial, gelatinous, thallus comprised of several intertwined and contorted monoliform filaments having heterocyst or big cells in midway of each filament.

Nostrils (Nares) Paired openings between the nasal cavity and the exterior invertebrates. Fish has only external nostrils, opening to the exterior but higher vertebrates also have internal nostrils opening into the buccal cavity.

Notonecta/Back Swimmer Insect belongs to order Hemiptera, Family-Notonectidae. Generally bluish in colour, swim jerkily, body deep, convex, dorsally and keeled ventrally with side sloping, hind leg long and fringed for swimming grow to 10 mm in length.

Novel Protein Foods High protein foods that are not derived from livestock product or that are animal in origin but converted into food for human or livestock by novel processes.

NRCCF National Research Centre on Cold Water Fisheries conducts research on cold water fisheries resources of the country. The Head quarter is at Nainital.

Nuoc-mam Fermented fish sauce clear amber in colour made by fermenting small sea or fresh water fish stacked in alternative layers of salt flavouring ingredients and spices added, the decanting of the digestible enzymes takes several months, also dried.

Nursery habitat Habitat possessing conditions critical for survival of fish larvae and fry.

Nursery Phase The culture of post larvae from the time of metamorphosis to the time they are stocked in the on-growing ponds or released into the wild.

Nursery Pond A pond which is especially prepared for rearing the spawn (hatchling) to fry stage, for a period of 15 days. The size may vary from 0.1 ha to 0.4ha. The depth of water column may be between 1.0 to 1.5 m.

Nutrient Budget A budget drawn up in terms of nutrients for a living system.

Nutrients It refers to elements required to support the growth of phytoplankton in the water. These include usually phosphates, nitrates and silicates but sometimes also the minor elements of sea water such as copper, manganese, cobalt, iron etc. are included in this term.

Nutritional value of Fish Fish is a good source of high quality proteins that play beneficial role in lipid and cholesterol metabolism. The importance of fish lipids is due to their high content of polyunsaturated fatty acids which are effective in reducing plasma cholesterol and beneficial to ischemic heart patients. Fish protein has a high proportion of myofibrillar proteins(67-75%) and low proportion of storma proteins (3-6%) in teleost and 6-12% in Elasmobranches. It is rich in almost all essential amino acids, especially lysine & methionine. The essential amino acid content of fish proteins is identical with livestock proteins but it has high degree of digestiability. It has got hypocholesterolemic effect.

Nylon A synthetic fibre belonging to the polyamide fibre group.

Nymphaea It is commonly known as white water lily, an emergent aquatic plant, stem creeping, rhizome root tuberous long deeply rooted leaves disk like, 2 inches to 2 feet in diameter, notched, borne on long petiole, flowers big rubbery.

O

Obligate Descriptive of an organism that requires a certain environmental condition or nutrient for growth or survival.

Obligate Aerobic An organism that will only grow in the presence of oxygen.

Obligate Anaerobic An organism that will only grow in absence of oxygen.

Obligate parasite An organism that can only survive as a parasite on or in another animal.

Oceanography Study of the physical characteristics of the sea.

Occipital Pertaining to the occiput, the hindmost region on the top of the head.

Ocellus A small round eyelike spot on the fin.

Oedema The accumulation of fluid in tissues.

Oedogonium It is phytoplankton belongs to family oedogoniumceae. It is filamentous, unbranched cells, linearly elongated with reticulated chloroplast.

Off Flavours It is due to the excessive abundance of algae produces. The actual flavours originate from the decomposition of organic matter by various fungi. The chemical compounds responsible are geosmin, and methylisoborneol.

OIE Office of International des Epizootics is an international veterinary organisation in France, established in 1924 comprising 110 member countries worldwide. Its main objective are to promote awareness of serious animal disease problems associated with trade in live animals and means for their control of occurence and progress of epizootics of listed disease and the methods that are applied for their control in individuals country. This information is disseminated through a monthly Bulletin and a yearly report on the animal health situations world wide. Its fish disease commission (FDC) dealt with fish disease and related matters, It produces an annual report on the main developments regarding current diseases, new pathogens, diagnostic and control methods world wide.

Oil Immersion A microscopic technique using special high powered objective lenses. A drop of immersion oil (Cedarwood oil) is placed on the coverslip of a microscope slide and the objective lens carefully lowered into it. The oil has the same refractive index as the lens glass and increases the resolving power obtainable by letting a wider angle of rays enter the objective lens.

Oligohaline Water that is less salty than sea water usually refers to the area where fresh water and sea water mix such as in an estuary, bay or areas adjacent to the shoreline.

Oligotrophic Aquatic systems with low nutrient concentrations and low levels of biological productivity.

Omnivore An organism that feeds on both plant and animal substance.

On-Growing Growth to market size.

Oocyst A one or three layered wall or membrane bound pack surrounding a sporont or sporocysts.

Oocyte The eggs cell contained in an ovary, which under goes meiosis to form an ova and a polar body (a cell with a little cytoplasm). The first division of meiosis occurs in a primary oocyte which generates the secondary oocyte. This then undergoes the secondary meiotic division to produce the ovum. In some species the cell released from the ovary and with which fertilization occurs may be at the oocyte stage.

Oostegite Thoracic plates that extend under a female isopod to form a marsupium in which the eggs are carried.

Oozing The fully ripe condition of the gonads in which the milt of the male as well as eggs of the female are ready to be shed.

Operational Risks Risks which interrupt the production cycle e.g mechanical failure, failure of technical processes, late delivery of supplies and services. It includes operational, technological, financial and social risks.

Operculum It is a large bony flap covered by skin, over the gills on each side. It is absent in cartilaginous fishes. In bony fishes, it is made up of opercular, preopercular, subopercular and interopercular bones.

Opportunistic Infections A disease caused by an organism that is not usually associated with the particular host.

Opportunity Cost The cost of taking an action in terms of foregone alternatives.

Optimum Population The number of individual that can be accommodated within the area to the maximum advantage of each of them. The concept may be interpreted ecologically or economically.

Optimum Take The annual catch from a given stock in a given year which will neither harm the reproductive potential of the stock, nor alter materially the aquatic ecosystem and is also economically feasible.

Optimum Yield Optimum yield in reality undefinable quantity used in some fisheries management considerations.

Oral vaccination The immunizaion of an animal in which the viccine is mixed with food.

Oral valves A pair of folds present in maxillary and mendibular regions of mouth, permit water to enter the buccal cavity but resist the back-flow, are known as oral valves.

Orange Claw Male Fast growing but sub-dominant male *Macrobrachium rosenbergii* distinguished by large orange claws.

Organic Farming/Biological Husbandry Farming without the use of artificial fertilizer or pesticides.

Organic Load The amount of dissolved organic matter carried in the water.

Organic Manure Manures containing organic matter, e.g. farm animal excrement. The following manures are often used in fish pond, faeces, urine of live stock, polutry, night soil, green manure, compost and silk worm dregs. Only through decomposition by microorganisms the organic manure is converted to nutrients that the plants can absorb.

Organic Soil It have lighter dry weight, colour is brown or dark brown water holding capacity is high, on volume basis a cubic foot of peat soil at optimum moisture can supply somewhat more water to plants than a representative mineral soil also at optimum. A organic soil has invariably a better physical condition. On the basis of their organic content soils of two kinds are recognised (i) mineral (ii) organic. The mineral soils vary in organic matter from a mere trace to as high as 12 to 15%. Organic soils possess much more organic material, the content which ranges from 20 to as high as 90-95% in some cases.

Organochlorine An organic compound containing chlorine, sometimes called chlorinated hydrocarbon. Many have biocidal properties and are used as the active ingredients for pesticides with a high persistence due to their chemical stability and low solubility in water, pesticide applications include DDT, aldrin and lindane. These have proved very effective in the control of insects pest. They persist in sublethal doses in animals, where they accumulate in fatty tissue. Low concentration of residues may be increased as they pass up the food chain and toxic levels may thus occur in the bodies of predators causing death or sterility. Because of these side effect the use of many of these insecticides has been limited in the UK and in other countries.

Oriental Cure Salted ungutted herring produced for oriental trade.

Orlon A synthetic fibre belonging to the acrylic fibre group.

Osmolarity The concentration of osmotically active particles in solution.

Osmoregulation Regulation of the concentration of dissolved substances in the body fluids (mainly by diffusion) through semipermeable tissues/membranes.

Osmotic Potential The maximum osmotic pressure that could develop due to osmosis in a solution separated from pure solvent by a semipermeable membrane.

Osmotic Pressure Pressure generated by materials or ions in solution results in an absorption or loss of water until any differences in concentrations are equalised.

Ostracod A small group of minor crustaceans.

Otolith One of the several granules of calcium carbonate that are connected in a gelatinous mass and attached to hair like processes of sensory cells within the utriculus and saculus of the vertebrate inner ear. They respond to changes in the position of the head and so stimulates the sensory cell.

Ottelia It is a floating aquatic plant. The rhizome root filamentous in bunch, leaves round or ribbon like borne on 10-20 cm long petiole.

Otter A common fish eating carnivore of aquatic habits.

Out Breeding Breeding between individuals that are not closely related, in animals behavioural mechanisms often promote outbreeding. The most extreme form crossing between species usually results in sterile offspring and there are various mechanisms to discourage it. Out breeding increases hetrozygosity giving more adaptable and more vigorous populations.

Ovaprim It is the trade name of induced breeding agent of fish. It contains analogue of salmon gonadotropin releasing hormone and domperidone, a dopamine antagonist both dissolved in caliberated quantities in a non toxic organic agent. This is manufactured by Syndel Laboratories Ltd. Canada.

Environmental Stimuli
↓
A Hypothalmus
↓
B Releasing Hormone
↓
Pituitary
↓
C Gonadotropins
↓
D Steroidogenic Ovariantissue

Estrogens → Vitellugenesis

17 alpha 20 Beta progesterone

↓
Mature eggs
↓
ovulated eggs

Lettered areas are points in the endocrinologic pathway where artificial application have successfully induced fish to full or partial ovulation ovaprim acts at point B in the Pathway.

Ovary A female reproductive organ that produces eggs in animals and megaspores in plants.

Over Fishing The depletion of a fishery resource to such an extent that a stable population can not be maintained through normal reproduction rates.

Overcast The state of the sky when more than nine-tenth of the visible canopy is covered with clouds.

Overturn The circulation of water and nutrients in a lake when the thermal strata become mixed by the wind during the spring and autumn.

Oviparous An animals that lays eggs.

Oviposition Release of eggs from the fish.

Ovoviviparity A property of some invertebrate, fish and reptiles involving the incubation of growing young within the body of the female. The eggs are fertilized internally and the developing embryos are nourished by large eggs rather than placental tissue. The young are hatched with-in or right after extrusion from the parent. Several species of shark exhibit this characteristic.

Ovoviviparous Species that incubate eggs internally but yolk reserves provide most or all of the energy available for embroylogical development.

Ovulation The release of an egg (ovum).

Ovum An unfertilized nonmotile female gamete produced in an ovary.

Ox-bow Lake or Cutoff A lake formed when a meandering river cuts a new course through the neck of a meander, isolating part of the old stream course creating a lake.

Oxidation The process of combining with oxygen or the removal of electrons from ion or atom.

Oxidation Pond A shallow lagoon or basin in which waste water is purified by sedimentation, aerobic and anaerobic treatment.

Oxygen Absorbed The amount of oxygen in parts per million of water abosored by these sample of water from acidic permanganate at 27°C in four hours.

Oxygen Cycle The cycle in which oxygen is taken in by animals and given off in the form of carbon dioxide, which is then taken in by plants and given off in the form of oxygen.

Oxygen Debt A physiological state that occurs when a normally aerobic animal is forced to respire anaerobically during a temporary shortage of oxygen (anoxia).

Oxygen Minimum Layer Usually observed in the tropical area is a sub surface layer, in which the dissolved oxygen content is very low or nil.

Oxygen Quotient The rate of oxygen consumption of an organism or tissue. It is usually expressed in microlitres of oxygen per milligram of dry weight per hour. Small organism stand to have higher oxygen quotient than larger ones.

Oxygen Sag Curve A line drawn on a graph to trace the dissolved oxygen content of water to purify itself following the discharge into it of an effluent.

Oxygen Saturation Capacity The maximum quantity of dissolved oxygen that a liquid exposed to the atmosphere can contain at a given temperature, pressure, and salinity.

Oxygen Tension The oxygen tension represents the partial pressure of oxygen in the atmospheric required to hold a certain concentration of oxygen in the water.

Oxygenation Addition of oxygen to water.

Oyster A gray bivalve mollusc with one convex and one flat shell. Edible oysters are native to almost all waters of Northern Hemisphere. It ranges in size from 7 cm to 30 cm. The young spat is a ciliated, planktonic swimmer for a few days after hatching. It then attaches itself to a rock and filters water for food. All oysters form pearls but only some pearls are lustrous.

Ozonation Addition of ozone to water to break down refractory organic molecules, oxidize waste metabolites and sterilize the water.

P

P/B A ratio of primary production to total biomass.

Pablo Boat It is a trawler used along the coasts of Andhra Pradesh, Tamil Nadu, Kerala. At present it is used in Lakshadweep.

Padda Fish product consisting of slices of fish that have been dipped in paste of clarified butter or oil with chillies mustard and other spices and packed in jars.

Paddel Wheel An electrically driven floating device used to aerate and circulate water in ponds.

Padec Fermented fish paste made with rice husks.

Pair Trawl Mid water or bottom trawl towed and held open horizontally by two vessels.

Panne Shallow depression in high salt marsh habitats that holds a permanent pool of water.

Paper Shell Recently moulted crustacean whose shell has started to harden and turned leathery.

Papillae Small conical projections on the body surface.

Papillon Trade name for salted cod of a weight less than 400 gm at the time of landing.

Parallel Filament Ropes These are continuous filament arranged parallel to each other and compressed into a cylindrical bundle of filament which is then covered by a plastic jacket which is extruded over the bundle. Because of their lower elongation they are suitable for replacing steel wire or combination ropes in fishing gears for instance as framing ropes for high opening mid water trawls, purse seine and gill net.

Paramecium The genus belong to the phylum-Protozoa, order-Holotricha. It is microscopic, unicellular organism, body slipper shaped, posterior end pointed body covered with small cilia.

Paraquat Organic contact herbicide used to control many broad leaved weeds and grasses. It is poisonous and accidental ingestion of large amounts, usually by children has caused a number of deaths but used correctly it is rapidly adsorbed on soil particles and so inactivated.

Parasite An organism in or on the body of other organisms deriving its sustenance from its host.

Parasitic Fish The small transparent South American cat fish *Stegophilus insidiosus* sucks blood. It is dangerous to man because it may penetrate the urethra while bathing. Removal can only be effected by surgery. Since barbs on the gills covered do not permit pulling it out.

Parasitism A close association between living organisms of two different species during which one partner, the parasite, obtains food at the expense of the other, the host. Many animals and a few flowering plants (dodder) are parasities. Pathogenic bacteria and fungi are parasitic on plants and animals and viruses parasitize plants, animals and bacteria.

Parasthesia Morbid or abnormal sensation such as burning prickling.

Paratenic host A transport host in which the larvae stage of a parasite undergoes no development and its only function is to transfer the parasite to the next host.

Parental care The practice of looking after the eggs and young one exhibited by some fishes and other animals.

Parr Young salmon before it leaves fresh water for the sea.

Particulate Organic Matter The weight loss after ignition at 550°C of the residue on the filter from the total particulate matter.

Passive Capture gear Stationary fishing gear that captures fishes when they contact it during their daily movement patterns.

Pasteurised Fish Fish and seafood packed in cans, glass jars or other containers absolutely airtight, preserved by heating at temperature below 100°C for a limited time, to be chill stored.

Pathogen Any organism or substance causing disease.

Pathogenic Waste Waste that could contain pathogens and whose disposal might therefore threaten public health. It includes hospital wastes, laboratory wastes, etc.

Pathognomonic Specifically distinctive or characteristies of a disease or pathological condition, a sign or indicant on which a diagnosis can be made.

Pathology The branch of medicine concerned with diseases, their causes, and their effects on the human body.

Pay Back Period It is the number of years required to recover the loans.

Peak Discharge A period of maximum stream flow when the base flow is augmented and surface run off is greatest.

Pearl A hard smooth lustrous protective structure formed around an irritant by the mantle of a shelled mollusc, whose shell interior is necreous (mother of pearl). The coating applied to the irritant consist mainly of calcium carbonate and some protein. A round or pearl forms in a part of the mollusc not subject to muscle pressure . If the pearl forms where it is affected muscle action it is deformed or flattened. Gem quality pearls are found in both marine and freshwater mollusks. They vary in colour from white to cream rose, lavender green gray and black, depending on the location and producing animals. A cultured pearl is a pearl formed in a captive oyster by inserting an irritant into the oyster and then maintaining it in a protected environment for several years.

Pearl Essence Liquid suspension of the particles of guanin, the lustrous material extracted from scales of fish such as herring, sardine or their swim bladders.

Pearl spot The brackish water perch *Etroplus surantensis.*

Peat Unconsolidated soil material consisting largely on decomposed or only slightly decomposed organic matter accumulated under condition of excessive moisture.

Peat Soil The colour of typical cultivated peat soil, dark brown or intensely black when it is wet, is perhaps the first physical characteristic that attarct the attention. It has a light weight also. It has high water holding capacity. A typical woody or fibrous peat soil is always having an ideal physical condition.

Pebble A rounded fragment of rock, it is larger than a gravel stone but smaller than a cobble. It is sometimes defined as having a diameter of 4-64 mm. The rounding of rock fragments may result from either wind or water action. Large numbers of pebbles that accumulate in one place form a pebble bed.

Pectoral fin The front or upper most of the paired fin at the level of the shoulder.

Pectoral spines Modified, stronger rays of the pectoral spines.

Pedigree The ancestral history of an individual, a chart showing such history.

Pelagic Fish that live in mid water.

Pellet Compounded feed ratios of graded sizes containing about 10% moisture.

Per Unit Area Production cost It is the cost indicator of a farming. It can be calculated as,

$$\frac{\text{Production cost of a product}}{\text{Area for aquaculture (Ha.)}}$$

Per Unit Product Output Value Cost It can be calculated as,

$$\frac{\text{Production cost of product}}{\text{Total output value of the product}}$$

Per Unit Production Cost It can be calculated as,

$$\frac{\text{Production cost of certain product}}{\text{Total yield of the product}}$$

Percolating Water Those water that pass through the ground beneath the earth's surface without a definite channel and not shown to supplied by a definite flowing stream.

Percolation The movement under hydrostatic pressure of water through the interstics of rock or soil.

Perennial A water bodies which not dry up completely during summer.

Perepods The walking legs of crustaceans commonly spelt pereiopods.

Periphyton The organisms attached to under water rooted plants or any other substance.

Permanent Eggs Numerous lower animals such as rotatoria, Daphnia etc. are able to survive in unfavourable periods, such as intense cold or drought by producing very large unfertilized eggs which are rich in yolk and are protected by rugged shells. These insure the survival of the species during such periods.

Pesticide Chemical agents, most often artificial that kills plants or animals. A general term embracing insecticides, herbicides, fungicides, nematicides, etc. Some pesticides e.g. DDT have had wide-spread disruptive effects among non-target species.

pH The measure of the acidity or alkalinity of a substance based on the number of hydrogen ion in a litre of the solution and expressed in terms of $p^H - \log_{10}\left(\frac{1}{H^+}\right)$. Where H^+ is the hydrogen ion concentration. Pure water dissociates slightly into hydrogen and hydroxyl ions $(H^+ + OH^-)$ the concentration of each type of ion being 10^{-7} mole per litre, so the pH of water is $\log_{10}\frac{1}{10^{-7}} = 7$ and this is taken to represent neutrality on the pH scale. Smaller values are acidic, larger values are alkaline.

Phenolphthalein An acid alkali indicator that is colourless in acids and red in alkali. It has a pH range from 8.4-10.0 and is a frequently used as indicator for the detection of pH change in acid alkali titrations.

Pheromones A chemical substance released by fish into the external medium which influences the behaviour of other recipient fish of the same species transmitted with information through olfactory receptors. Pheromones are intraspecific semi-chemicals . The transmission signals produced by an animal/fish and semi-chemical in nature and are known as allomones. The pheromones may be proteins or amino sugar or sex-steroids.

Photic Zone The surface of greatest depth at which sufficient sunlight penetrates the water column to support photosynthetic organisms.

Photodegradeable A description of somethings that decomposes under the action of light.

Photoperiod/Photoperiodicity The response of plants or animals to the relative duration of day and night. The timing of the breeding season in many vertebrates and the flowering period of many plants are regulated by day length. The concentration of planktons also vary with day and night in the ponds.

Phototactic Swims or moves towards light.

Physical Hazards The risk or danger arising from the subject matter of insurance for example, excessive flooding when the crop is about to be harvested.

Phytoplankton A general term for freely floating mainly microscopic plankton algae of the water.

Pickle Cured Fish Fish that have been treated with salt in water tight container so that the fish are cured in the pickle that is formed.

Pig Manure Pig manure consist of organic matter and other nutritional elements such as phosphorus, nitrogen, potassium. The pig faeces are delicate containing nitrogen (C : N = 14 : 1) making them more susceptible to rotting. The major portion of pig urine is nitrogen in form of urea.

The nutritional elements in Pig manure is :

	Organic matter %	*N*	*P_2O5*	*K_2O*
Faeces	15	0.6	0.5	0.4
Urine	2.5	0.4	0:1	0.7

The excretory amount of a pig is associated with its body weight and food intake. A 50 kg pig discharges about 10 kg/day or 20% of its body weight per day. A pig excretes 500 kg of faeces and urine in growing period of 8 months. from piglets to adult.

Pinnipeds Group of marine mammals that includes seals, sealious, walruses.

Pimozide Drug used to induce spawning artificially works by stimulating gonadotropins-release.

Pineal Gland A gland in the upper region of fish brain which can sense light.

Piscivorous The animals which eats fish as food.

Pistia It is a floating aquatic plant commonly known as water lettuce. The stem is tubular creeping, root fibrous, adventitious, leaves sessile, entire with silky texture, flowers in group.

Pitch (of rope) Distance between two successive spirals of the same strand of a rope measured parallel to the axis of the rope.

Pitechial haemorrhage A tiny haemorrhage about the size of a pin head.

Pituitary Extract Aqueous, alcoholic or acetone dissolved extract of pituitary gland used for hypophysation or the artificial induction of spawning.

Pituitary Gland Endocrine or hormone producing gland found on the upper side of the brain just behind the eye.

Placenta in scoliodon Scoliodon is viviparous. Usually 4-6 embryos develop within the uteri and are born alive. The embryo is first nourished by yolk contained in the yolk-sac, connected with the intestine of the embryo by means of a yolk stalk. When all yolk is used up, the yolk-sac shrivels up and its connection with the intestine is closed, but the yolk- sac in the mean time becomes embedded in the wall of the uterns. Arterial and venous extensions grow into the yolk-sac through the yolk-stalk and nourishment is absorbed directly from the uterine wall by way of diffusion in the blood capillaries. Thus the yolk-sac is converted into a placenta, and the yolkstalk into the placental cord. Numerous appendicula grow out from the placental cord, probably they aid in absorbing nutrient material from the uterine walls.

Placoid scales The characteristic structure of the elasmobranch integument is the presence of placoid scales (dermal denticles) These scales are arranged in oblique rows, cover the entire surface of the body and impart roughness to the skin. A typical placoid scale Consists of a backwardly directed trident spine mounted over a rhomboid basal plate. Basal Plate is perforated in the middle and below for the passage of blood vessels, connective-tissue fibres, and nerves into the hallow interior of the trident spine called the pulp-cavity. Surface of the spine appears to have numerous plates arranged like tiles in an imbricating fashion. Basal plate remains embeded in the dermis and is composed of a hard substance like bone. Spine which cuts through the epidermis and projects over the surface,is largely composed of hard dentine reinforced by a layer of a harder substance called vitrodentine.

Plankton Microscopic organisms that are suspended within the water column usually drifting passively with currents.

Planning It may be defined as the way to attain an objective or goal. It may be divided simply into three steps, defining a problem or a need, identifying the resources that are required to solve the problem or fulfil the need including an assessment of the best alternative among the options, available and preparation of a plan of action and a budget to provide the required resources.

Plara Headed, gutted fish or pieces of salted and then fermented.

Pleopod The abdominal paddles or swimmer of a crustacean, also used to provide attachment for incubation of eggs.

Pleurodont In pikes and anglers teeth are hinged so that they yield to the pressure and snape back when it is over. In certain other fishes teeth are fixed in a groove curved on the inner surface of jaw bones, and are fused with jaw bones, such a condition is called pleurodonty.

Ply Number of yarn in a strand or total number of yarns in a twine.

Pneumatic duct The duct which connects the swim-bladder with the digestive tract is known as pneumatic duct. In most higher bony fishes, the pneumatic duct is lost and the air-bladder is like a closed sac. In such fishes air-bladder has no connection with pharynx, such fishes are known as physoclistous, *e.g*, Perch,cod. Fishes with a pneunmatic duct opening into the Oesophagus are called physostomous, e.g. carps, eel, *Acipenser, Amia*, Salmon, *Polypterus, Protopterns, Lepidosiren, Ceratodus.*

Poikilosmotic The term applied to animals (e.g. marine invertebrates) which body fluids with an osmotic pressure varies according to that of the surrounding water.

Poikilothermic The inability to control the internal body temperature hence changes passively with ambient condition.

Point Cut Cutting of two legs in each knot. All point cut gives a right angle webbing.

Pole and Line Fishing A fishing method using pole line with hook.

Pole Scraper This is an instrument to scrape the sampling of underwater growth on poles, embankments etc. This have a facility for sweeping the underwater poles by the help of metal scratcher or brush attached to it.

Pollutants Substances that have a harmful effect on living things.

Polyculture Farming of different fish species (that eat different foods) together.

Polyester A synthetic fibre of density 1.38g/cm^3 and melting point 250°–266°C. In this the fibre forming substance is a long chain of synthetic polymer composed of atleast 85% by weight of an ester of glycol and terepthalic acid. Some wellknown trade names are Terylene, Dacron etc.

Polygonum It is commonly known as water smart weed, stem long, erect and branched, root woody, and creeping, leaves lanceolate oblong or cordate arranged alternately, flowers pink in colour and arranged on the terminal spike.

Polyphyodont In the sharks and rays, teeth are not fixed over the jaw-cartilages but are carried on folds of skin or tough mucous membrane covering the jaws. Because of such "loose and weak" attachment teeth are likely to break off, therefore, they are continually replaced as they are worn out or broken off such a condition is called polyphyodonty and the dentition is called polyphyodont.

Polypropylene Synthetic fibre of density 0.91 gm/cm^3 and melting point 160°–175°C. The fibre forming substance is a long chain synthetic polymer composed of atleast 85% by weight or propylene.

Polysporic A plasmodium which produces several spores.

Polythelene Synthetic fibre of density 0.96g/cm^3 and melting point 125°–140°C. The fibre forming substances is the long chain synthetic polymer composed of atleast 85% by wt. of ethylene.

Pool Section of a stream or river where some portion of the water is not noticeably in motion.

Population Group of organisms that interbreeds freely among themselves but is isolated often due to geographic barriers from interbreeding with other groups of same species.

Post Larva The stage following the last planktonic larval stage in crustaceans, Frequently the time of transition between planktonic and benthic existence.

Potamageton It is a submerged aquatic plant, the stem branched, long, slender, soft, height varies with depth, root filamentous, leaves small to short, ribbon like, margin serrated, smooth and wavy.

Poultry Manure It includes the faeces of chicken, ducks, goose and are rich in organic and inorganic matter. Poultry manure rot quickly and their nitrogen is mostly in the form of uric acid. The annual excrement of per fowl is, chicken- 5.0 – 5.7 Kg, duck 7.5 – 10 Kg, goose 12.5 – 15 Kg. The nutritional elements in poultry manure is,

	Organic carbon %	*N*	P_2O_5	K_2O
Chicken	25.5	1.63	1.54	0.83
Duck	26.2	1.14	1.44	0.62
goose	23.4	0.55	0.50	0.95

Pox Disease/Papilloma Disease Dense formation of papilloma in various sites on trunk head and fins in carp and gold fish. It is white or pink and have a flat gloss appears like wax in first sight. The shape and expense varies but the area ranges from 1 cm^2 to 7 cm^2 or even broader. The growths are difficult to peel off and hemorraging occurs when done forcibly. The growth of the fish is hampered become weak and die due to the secondary infection.The causative agent is virus.

PPM Parts per million A unit of chemical measurement used for reporting the level of trace materials (e.g. oxygen dissolved in water) Or of a additives (e.g. Formaline or chlorax). It is equivalent to 1ml/m3, 1gm./ton, or 1 mg/litre. Example-1. (for liquid) Add 50 ppm formaline in 250 litre of water. The expression 50 ppm means 50 parts of formaline to every 1 million parts of water. As 250 litres = 250000 ml thus amount to add (in ml.) is,

$$\frac{50 \times 250{,}000}{1000000} = 12.5 \text{ ml}$$

Thus 12.5 ml of formaline must be added to 250 litres of water to provide a concentration of 50 ppm.

Example 2 (for solid) Add 4 ppm copper sulphate and volume of water is 1.5 m^3. Then the amount of copper sulphate to add should be,

$$\frac{1.5 \times 1000 \times 1000 \times 0.4}{1000000} = 0.6 \text{ gm.}$$

PPT (0/00) Parts per thousand A unit of measurement usually applied to salinity. It is also expressed as 0/00.12°/00 water means the salinity of 12 ppt. It indicates that in 1000 ml of this water 12 gram salt is available. The salinity of full sea water varies but is normally around 35°/00.

Prawn Crustacean, taxonomically equivalent to shrimp but in use the term differs between countries. Prawn is extensively described as Macrobrachium species while shrimp refers to Penaeids.

Pre-operating Expenses Cost of feasibility studies, construction and legislation requirements.

Prebiotics These represent an attempt to achieve the same effect with a nonviable supplement the substances used (Oligosaccharides) stimulates the beneficial components of the gut, microflora and may be used along or in combination with probiotics.

Precocious Maturation/Precocity Maturation at an earlier age than normal.

Predator An animal living by preying on other animals.

Prefeasibility Study A preliminary assessment of a culture proposal designed to determine the need for a scope of a full feasibility study.

Preformed The wires and strands shaped to the helical form they assume in the strand and rope.

Prefuran It is a highly strategic antibiotic and antifungal drug which can be used alone. It is absorbed rapidly across the gill space. It is used for treatment of animals that are not feeding due to bacterial rot in the post larval stage.

Premix A mixture of vitamins, minerals and other micro-ingredients which is added to a compounded diet during manufacturing.

Prepatent period The period after infection but before the causative agent can be found using the usual diagnostic techniques.

Preservation of Net Treatment of the net with preservatives to reduce deteriorations caused by rotting, mechanical wear and tear heating etc.

Press Cake The residue after the pressing stage in the manufacturing of fish meal and oil from fatty fish, contains about 60% water and 4 to 5% oil. It is usually further dried and ground to make meal.

Prevalence The percentage of animals in a population which are infected at any one time by particular organism.

Primary production The production of biomess by plants through photosynthesis.

Primary Productivity The development of diverse aquatic communities based on Phytoplankton growth in culture and may be enhanced by the addition of fertilizer.

Primary Sexual Characteristics All sexual characteristics which are directly related to the sexual function such as ovaries, testis etc. In fishes generally it is located in the interior of the body, so that one can not readily make use of them for identifying the sexes.

Probiotics A live microbial feed supplement which beneficially affects host animal by improving its intestinal microbial balance. Prominent Probiotic bacteria is Lactobacilli, which are also important in producing organic acids which have an antibacterial effect. Probiotics are currently produced in the form of powders tablets, pastes, oil suspensions and sprays. The type of preparation used will depend on the animal being dosed and the effect required but by far the easiest and most commonly used form of administration is by inclusion in the feed or water. The benefit of Probiotics may be summarised as, (i) Promotes growth and Productivity (ii) Protect against salmonella infection (iii) Protects against toxins produced by harmful form of E.coli (iv) It stimulates the immunity to infections by boosting interferon production immunoglobulin concentration and macrophage activity.

Procercoid The first larval stage of many cestodes which develop inside the body cavity of the invertebrate (first) intermediate host.

Production The creation of biomass within a population through reproduction and body growth

Production (Fish) *Value per unit area (Rs/ha)* : It can be calculated as,

$$\frac{\text{Total fish production (Rs.)} - \text{Total Production cost (Rs.)}}{\text{Total pond area (ha)}}$$

Production Cost Production cost is the sum of the total expenses for producing per unit of fish, crop and animal products etc. It can be calculated on the basis of the total production cost of the farm or the cost of each activity.

Fish production cost per unit area (Rs/ha) : it can be calculated as,

$$\frac{\text{Total Production cost of fish farming (Kg.)}}{\text{Total area or the fish pond (ha)}}$$

Total production cost per unit area of the farm (Kg/kg) : it is,

$$\frac{\text{Total sum of production costs of all activity in the farm}}{\text{Total area for the integrated fish farming system.}}$$

Production Efficiency Average cost is as low as possible for a given level of output shows the production efficiency.

Production Function It shows the relation between inputs and outputs. It is often stated as a mathematical equation.

Production indicators Production indicators mainly reflect the productivity and the level of technical and economic management. It is also known as yield indicator.

Production Pond The pond used for rearing the fingerlings up to market able size fish. The size of the pond varies from 0.4 to 2 ha, as ponds larger than this are unsuitable for efficient management. The depth of water column may be between 2 to 3 meter.

Production Value/Yield Value It refers to the yield or production value per unit area activity or the total production value of a farm. The formula for the calculation is as follows,

Fish yield (Kg/ha) or fish production value (Rs) per unit area,

$$\frac{\text{Total fish production (Kg.) or value (Rs.)}}{\text{Total pond area (ha.)}}$$

Productive inflammation see prolefirative inflammation.

Productivity The amount of output resulting from the use of inputs. Productivity can be stated as out-put per unit of input (average product) or the increased out put resulting from an additional unit of input (marginal product).

Professional Breeders The term coined for the brood fish which undergo for multiple spawning in a year.

Profit It is the difference between the gross revenue and the total annual operating cost of the production.

Profoundal Zone The portion of water column under the pelagic Zone and immediately above the substrate. The substrate does not support rooted aquatic vegetation in the profundal.

Progenesis Larval stages which attain sexual maturity and produces eggs.

Project A plan with a purpose. It can be defined as, a project as an scheme or part of a scheme for investigating resources which can reasonably be analysed and evaluated as an independent unit.

Prolefirative inflammation An inflammation which is characterized by a pronounced multiplication of fibroblasts histoocysts and endothelial cells.

Prolific With high reproductive capacity.

Propeller /Aspiration Pump An electrically powered floating device for drawing air into a stream of water discharged at an angle towards the pond bottom, serves the similar function as paddle wheel.

Protandry A type of hermaphroditism in which the individual first function as a male and then as a female.

Protective food Food which contains the essential requirements for body building.

Protein/Energy Utilization Rate (%) It represents the relationship between input (Feed, fertilizers) and out put (fish, livestock) and is used in appraising the economic returns. It can be calculated as,

$$\frac{\text{Total amount of protein or energy of produce} \times 100}{\text{Total amount of protein or energy of input.}}$$

Protocercal caudal fin It is the most primitive type of fin. It occurs in Dipnoi (lung-fishes) and in the developmental stages of all fishes. In this fin, the vertebral column is straight and the tail is divided into two equal parts or lobes. Thus, the upper or dorsal epichordal lobe and the lower or ventral hypochordal lobes are equal.

Protogynous Type of hermaphroditism in which all individuals are female upon reaching sexual maturity and may become males at some later time.

Pseudocyst A cyst like structiure surrounded by a dense fibrous capsule. It does not possess an inner lining and the formation of pseudocyst follows necrotic changes.

Pseudomonas Infection The most typical external symptoms are superficial shallow greyish red circular or irregular ulcers. The mucous membranes and cartilagenous tissue in and around the mouth may be inflamed and even eroded. Fish may have distended abdomen which is filled with slightly opaque or bloody fluid, exophthalmus, fish may be inflamed.

PUD Peeled and un deveined.

Pump Fishing with Light Pumps are used to suck the fish from purse seine or directly from water. In the latter case the fishes are attracted first by light.

Pure Risks The risks of life and bussiness in general. It includes the physical risks of nature, social and political risks and liability.

Purge The removal of unsightly gut contents by holding the crustaceans without food in clean water for one to two days.

Purse Cable It is also called purse line wire rope employed to close the bottom of the purse seine.

Purse Seine Net Net that surrounds and traps large schools of fish near surface.

Pyloric caeca Many fishes have pyloric caeca, give off from the portion of the mid-gut, lying immediatley behind the pylorns. These caeca may be blind, tubular or sac like diverticulas of uncertain functions. They do not flow any definite arrangement or the organisation and vary in numbers in different fishes. Only a single such caecum is found in Polypterus while the number may go upto two hundred in mackerel.

Q

Quality The totality of features and characteristics of a product or service that bear on its ability to satisfy stated or implied needs.

Quality Audit A systematic and independent examination to determine whether quality activities and related results comply with planned arrangements and whether these arrangements are implemented effectively and are suitable to achieve objectives.

Quality Management The overall management function that determines and implements the quality policy.

Quality Policy The overall quality intentions and directions of an organisation as regards quality as formally expressed by top management.

Quality System The organisational structure responsibilities procedure for implementing quality management.

Quanternary Ammonium Chemicals used in the treatment of external bacterial or parasitic disease.

Quarantine Enforced isolation of organisms which are or may be infected to prevent transmission of diseases to other organisms or the environment.

Quick Lime/Calcium Oxide/Unslaked Lime A white solid formed by heating calcium in oxygen or more widely by the thermal decomposition of calcium carbonate. On a large scale lime stone is heated in a tall tower called lime kiln to a temperature of 550°C.

Quinaldine It is an anesthetic used for fish transport. It is a toxic solution and must be applied with care. It is worth using only for mass transport in big tanks. Effective concentration ranges between 10-30 mg/l according to species.

Quota Level of harvest that cannot be exceeded by fishing fleets participating in the fishing.

R

Race Group of members of a species which differ from their fellow members by some characteristic but are able to inter breed and produce fertile offsprings.

Raceway Elongated fish rearing facility usually fabricated from concrete, water enters at one end and leaves from the other.

Raft Rafts are commonly used in inland waters. Rafts fitted over empty sealed barrels are useful in the inshore waters.

Ramie Bark fibre obtained from the plant, *Bochmeria nevia*. It is grown in Indonesia, Africa and India.

Rampani Net It is a shore seine which was first introduced along the South Canara coast. The most significant feature of this net is its considerable length which may measure upto 300 fathoms. The Rampani net is operated by about 60-80 fishermen.

Ranatra/Water Stick-Insect Insect belong to order Hemiptera and family Nepidae. Body is dull green in colour prothorax narrow than head, forelegs long and modified for grasping the prey, two long filamentous respiratory tubes at the posterior end grow to 20-40 mm in length exclusive of tracheal tube.

Rancid The breakdown of lipids and their constituents through poor storage condition which severely reduces their dietary value.

Random Sample A sample obtained in such a way that each individual in the population being studied has the chance of being selected.

Rate of Return A ratio equal to the annual income returned to the owner divided by the amount of investment.

Rate of Return on Annual Operating Cost It is obtained by dividing the profit by total operating cost.

Rearing Pond Pond used for rearing, advanced fry (2.5–4 cm) to fingerling stage (4-10 cm) for a period of two months. The size of pond is fairly larger than nursery pond, the depth of the water column may be between 1.5 and 2.0 m.

Recruitment The addition of new individuals into some life stage or size range of a population. Most often recruitment is referenced to sexual maturity (that is recruitment into the spawning stock) or to the size range that is vulnerable to fishing gear used in a specific fishery (recruitment to a fishery).

Red Diseases in Prawn All the stages of *P. monodon* are affected by the disease due to presence of aflatoxin in feeds, high pH of water, prolonged low salinity. It occurs mostly in summer and autumn, there is sudden drop in feed consumption in affected individual, lethargic, general body weakness; yellowish and eventually reddish discoloration of the body, short streaks of gills, reddish colour of faecal matter, poor growth, increased fluid in caphalothorax, emitting foul odour. Necrosis of hepatopancreas, haemocytic infiltration and fibrosis occurs after three weeks with gradual mortality.

Redox-Cline/Oxic/Anoxic Boundary/Exycline The penetration of oxygen into the soil reaches a certain depth where it forms the anaerobic border line. This coincides with a change in chemical conditions switching from an oxidizing to a reducing environment. This border is called Redox cline.

Reeving Setting a wall net on its head line by spreading the meshes evenly along the head line and hanging at regular intervals.

Refactory Organic Compound Dissolved organic matter not readily oxidized or broken down by biological filtration.

Refrigerated Food Finished foods that are not shelf, stable, which require storage under refrigerated conditions.

Regressive changes A pathological process which is related to metabolic disorders and it includes dystrophic changes and necrosis.

Releasing Hormone Peptide hormone, which controls all aspects of the pituitary gland.

Repair A process to reestablish anatomical and functional integrity of tissues or structure after an injury or inflammation.

Reproduction The formation of new individuals from parent it may be sexual when there is no need of gametes e.g. mostly in plants (vegetative reproduction) and sexual reporduction where gamelis (male & female) are formed.

Reproductive Capacity Reproductive capacity refers to the ability of fish producing their offsprings. The assessment of reproduction capacity is based on maturity age, sexual cycle, fecundity or amount of the mature germ cells, effective egg production, fry survival rate etc.

Resistance Hereditary strength to withstand diseases specialized to specific disease, or of general nature, generally developed to different degrees individually. Frequently only single individuals, though often also whole tribes are resistant to a certain disease. Due to this fact it is perfectly possible to breed resistant tribes. In order to become effective though hereditary resistance to be aided and developed through favourable living conditions.

Respiration The acquisition of energy from the oxidation of organic molecules in living cells. During aerobic respiration oxygen and food are consumed and carbondioxide and ammonia are produced.

Respiratory Quotient (RQ) The ratio of the volume of carbondioxide expired by an organism compared to the volume of oxygen consumed during the same period of respiration. A theoretical RQ can be calculated for the various food stuffs used in respiration giving a value of 1 for carbohydrates, 0.7 for fats, and 0.8 for proteins. However in practice more than one foodstuff is respired at one time and other metabolic processes may produce carbondioxide or use oxygen so an RQ measurement for an organism gives unreliable information about the type of foodstuff respired.

Resultant Linear Density (R—tex) Linear density of the product resulting from twisting, folding or cabling operations.

Return on Investment This is an indicator of investment analysis. It can be calculated as, gross investment divided by an average profit per annum.

Rhizoids Modified cytoplasmic projections which arise from the basal end of some parasitic dinoflagellates *eg.* Amyloodium; crepidoodinium these projections attach to or penetrate host cells to provide anchorage.

Riffles Roaches of a stream where the surface of the water is obviously turbulent rather than flowing in a laminar manner.

Rig The way of assembled the component of fishing gear.

Rigor Mortis Death stiffening.

Riparian zone Habitats occuring along the banks of river systems or the shorelines of lakes.

Risk Relates to an event which may or may not occur, the likelihood of which can be predicted with a certain degree of confidence.

RLG Relative length of gut is length of gut divided by total length of body. In general the length of gut in carnivores, herbivores, and microphages are one half, two and five times respectively of their body length while that of omnivores fishes varies between two and six times of their body length. The relative length of gut may change within the species with the change in their feeding habits. It is also related to the duration of intestinal digestion. The RLG of fish species having different feeding habits.

Fish species	*RLG*
Carnivores	
Channa maculatus	0.57
Anquilla Japonica	0.46
Clarias fuscus	0.68
Herbivores	
Ctenopharyngodon idella	2.16

Microphages	
Hypophthalmicthys molitrix	5.28
Sarotherodon mossambicus	6.29
Omnivores	
Cyprinus carpio	2.04
Carassius auratus	5.15

Rope Profing A chemical treatment given to natural fibre ropes for protection against microbiological attack and destructive agencies such as sunlight and sea water.

Rope A cordage with a circumference of 25 mm or more.

Rostrum The forward pointing spine between the eyes of a crustacean.

Rotenone Selective fish poison often applied in the form of dry derris root which contains about 5% active ingredient.

Rotifers A group of microscopic crustaceans which anterior end bears tuftes of cilia used for feeding and locomotion. Important living food resources for larvae frequently cultured in crustacean hatcheries.

Rotskjae Fish that has been split in two except for a short section of the tail, preparatory to hanging for natural air drying (Norway).

Rouelles Thin slices of fish cut perpendicularly to the back. bone applied mainly to mackerel (France).

Row It is a series of loops or half meshes aligned in the direction of braiding. It is half mesh in height.

Rudimentary Organs Organs that have been stunted in the course of time the reminders of which are still present today but have become purposeless.

Run Off That portion of the precipitation of an area which is discharged from the area through stream channels. That which is lost without entering the soil is called surface runoff and that which enters the soil before reaching the stream is called ground water runoff or seepage flow from ground water.

Runnage Length of twine per unit weight expressed as metre per kg, yards per lb. feet per lb.

S

S-1 Smolt or sea going stage of salmon which is transferred to sea water for on growing after one year in freshwater.

S-2 Smolt or sea going stage of salmon which is transferred to sea water for on growing after two years in freshwater.

Safe Working Load The maximum load which should be applied to a rope. It is always considerably lower than the breaking strength.

Safety factor The factor by which the breaking strength of a rope must be divided in order to obtain its safe working load. The factor may be between 4 to 12 according to the purpose for which the rope is intended.

Sagittaria It is a marginal plant, stem stoloniferous, rhizome roots fibrous, leaves petiolated, sagittate and vary in size, and shape.

Salinity A measurement of the total mineral ions into water expressed in ppt.

Salted Cured Fish Fish preserved or brine may or may not be dried afterwards.

Sample A selection of individuals drawn from a population.

Sampling The operation of collecting a sample from a population/water.

Sand Sand is a loose and single grained. The individual grains can readily be seen or felt. If squeezed in the hand when dried it will fall apart when the pressure is released. If squeeze when moist it will form a cast but will crumble when touched, will not form ribbon between thumb and finger.

Sandy Loam The soil containing much sand but which has enough silt and clay to make it somewhat coherent. The individual sand grains can be readily seen and felt. If squeezed when dry it will form a cast which will readily fall apart but squeezed when moist a cast can be formed that will bear careful handling without breaking. Sands and sandy loams are classed as course, medium fine or very fine depending on the proportion of the different sized sand particles that are present.

Sanitary Fish Herbivorous fish introduced into a culture pond to limit the growth of plants.

Saponin It is a hemolytic toxin ($C_{32} H_{54}O_{18}$) commonly found in the tea seed cake. At a concentration of 10 ppm it causes fish to die in 11 hours.

Saturation Value Maximum solubility of a chemical or gas in water, values increases with increased pressure and with reduced temperature and salinity.

Scale carp A variety of *cyprinus carpio* having scales covering the entire body surface.

Scales Scales are thin, flat . semi-transparent horny plates covering the body in most fishes. In the eels, they are visibly lacking though present deep into the dermis. Scales are completely lacking in the electric rays and cat fishes.

School A form of keeping together of fish communities. One speaks of a school when a number of fishes of same age swim together.

SCICI Shipping Credit and Invesment Company of India Ltd. A specialised development finance institution. It has specialised functional roles in the Maritime areas comprising broadly shipping and ship building, deep sea fishing, related equipment and infrastructure. It is a nascent body having been incorporated in December 1986.

Scoliosis Abnormal development or bending of the spine caused by diet or disease.

Scotch Cured Herring Fresh Herring free from feed, unwashed, gibbed roused packed tightly in barrels, mild cured in their own blood pickle, not repacked, limited keeping quality, pack about 700 fish to a barrel of 250 lb. net capacity.

Scrap Fish Non commercial fish, fish not used for human consumption.

Scratcher Disease Diseases of the intestinal canal caused by parasitic worms (Acanthocephala) which may be lethal to smaller fishes. The worms measuring fractions of an inch, dig into the wall of the intestine by means of their clasping organs, causing heavy inflammations perforations in the intestine, anaemia and weakening .

Scyphidia Small unshaped parasites which attach themselves to the body, gills and fins. Cilia at their top, beat activily

Sea Food Cocktail Product prepared from shell fish crustacean meal with sauces based on tomato ketchup seasoning etc. bottled unprocessed or pasteurised.

Sea Weed Meal Raw material for animal feeding stuffs prepared from dried brown algae particularly from *Ascophyllum nodosum, Fucus serratus, Fucus vesiculosis.*

Seasonal A pond which contains water for a part of the year, usually for four to eight months.

Secchi Dish A circular plate of 20 cm in diameter, the upper surface of which is divided into four quadrants painted alternately black and white. When lowered into a pond on a graduated rope the point at which it disappears provides a measure of turbidity or plankton density. It indicates the productivity of the pond. The ponds has been divided in five types based on the transparency of the secchi dish.

Secondary Pathogens Organisms whose infection of host is secondary to other factors in the disease causing process.

Secondary Sexual Character Sex specific characters which develop during maturation and are used for attraction or courtship help in activity. The following symptoms are indicative to ready to spawn in fishes :

Female

(i) Well rounded and soft abdomen, the fullness of which extends posteriorly past the pelvis to the genital opening.

(ii) The genital opening is swollen protruding and reddish or rose in colour its edge is uneven or fringed.

(iii) The anus may also be swollen and reddish.

Male

(i) The male releases a few drops of thick milt when its abdomen is pressed slightly.

(ii) The dorsal surface of the pectoral fin becomes rough.

Seepage Ponds Seepage ponds can be constructed in the depressed areas of ground in the delta of a river. These ponds are usually dugout ponds without the possibility of gravity drainage. This is serious limitation of this type of pond.

Seine The method of capture is by surrounding certain area and towing the gear, over this area with both ends to a fixed point either to the shove or on a vessel.

Seine Net / Seining Long fishing net which is suspended in the water by floats and is drawn through the pond to encircle the catch.

Sekiyama It is the third part of the branch line of tuna long line gear. It measures about 6 m. and is connected to the swivel on one end and to the wire loader on the other. It serves as an additional leader wire which avoids kinking maintaining leader wire straight so that the extanglement of the hook with the branch line is prevented.

Semiparity Life history in which all individuals of a population spawn during one spawning season only then die.

Semi-Intensive Farming This management is wide spread and here fish receives supplementary feeding.

Senescent The depression of cellular functions in an animal as part of the aging process.

Septicaemia Severe microbial infection which involves the presence of the infections agent circulating in the blood.

Sessile Term for organisms that live attached to a substrate.

Settleable Solids The volume or weight of solid material which will settle during one hour to the bottom of a Imhoff cone containing a measured volume of water.

Settling Ponds It is the pond specially designed for settling of the heavily loaded particles in the water before it enters to farm.

Shackle A staple like link closed with one of a variety of pins.

Sharp Frozen Fish Fish that has been frozen usually by laying it out in a low temperature cold store or on refrigerated shelves, has no precise definition and its use should be discouraged.

Sheave A grooved pulley for use with rope.

Sheep Foot Roller Heavy toothed roller for compacting clay soils during pond construction.

Sheep Shank It consists of two long bights in the rope and a half hitch over the end of each bight made in the standing part of the rope.

Shelf Stowage Stowage at sea of white fish laid side by side and head to tail belly downwards in single layer on 5 to 10 cm of ice but with no ice among and on top of the fish.

Shell disease Degrading of shell of larvae, post larvae, juveniles and adults of, *P. monodon P. indicus* and *P. mergusiensis* by bacteria belonging to *Vibrio, Aeromonas* and *Pseudomonas* causing brownish to black erosion of carapace, abdominal segments, rostrum tail, gills and appendages, development of blister containing cyanotic gelatinous fluid are the main symptoms of the disease, the affected shrimp becomes susceptible to cannibalism or dies from stress or energy exhaustion.

Shell Fish Term embracing both aquatic molluscs and curstaceans.

Shell On Processing term meaning with the outer skin (exoskeleton) left in place.

Shell valve One of the parts of the myxosporean spore wall.

Shidal Butki Sundried fish immersed in water drained and packed with fish oil in containers which are buried in the ground for several months.

Shiokara Fermented fish product made from squid or guts of skipjack, also other species brown salty viscous paste made by fermenting the raw material with salt in containers for upto a month, product packed in glass or plastic containers. (Japan).

Shoe Dhoni It is a popular craft found at Kakinada in A.P. It is wide and flat forward and narrow at the aft. They are driven by long bamboo poles in shallow beaches and canals some of these are provided with out board motors. Thcy are essentlally estuarine boats which operate stake nets.

Shoot Carp The asymmetrical increase in weight variation in a carp population due to interactions between individuals in the population and resulting from competition for food.

Short Seasonal A pond which will contain water only for 3-4 months in a year.

Shot Indicates a single piece of net in a fleet of gill net.

Shoulder spot A pigment spot near the pectoral or shoulder region seen in some Fishes.

Shrinkage It is the reduction in length of a specimen induced by wetting or other treatment e.g. setting.

Shucking A procedure in which a molluse is removed from its shell.

Sida Belongs to phylum arthropoda order-cladocera family sididae. The head large and separated from the body by a depression, rostrum pointed and forms a beak, a large cervical gland on head, 1st antennae jointed with a long terminal flagellum.

Side Piece/Side Wedge An identical piece of webbing attached on either side of belly to join the upper and lower portions of a four seam trawl. The portion of the webbing that comes above the belly is termed top wedge.

Silk Worm Dregs Silk worm dregs are composed of silkworm faeces, slough and mulberry residues. They are rich organic matter, dried dregs have 87 percent organic matter and 13 percent nitrogen. It makes good food for fish, 8 kg silkworm dregs produces 1 kg fish.

Silt Finely divided particles that are the intermediate form of particulate matter between sand and clay. Particles of silt range from 0.01 to 0.05 mm in diameter. Mud is within this range.

Silt Loam It has a moderate amount of fine grades of sand and only a small amount of clay. Over half of the particles being the size called silt when dry it may appear quite cloddy but the lumps can be readily broken and when pulverized it feels soft, smooth and floury. When wet the soil readily runs together and puddles. Either dry or moist it will form casts that can be freely handled without breaking, it will not form a ribbon if squeezed between the thumb and finger when moist but will give a broken appearance.

Silty Clay Loam It is a fine textured soil which breaks into clods and lumps that are hard to break with the squeeze of the hand when the clods are dry. When the moist soil is pinched between thumb and finger it will form a cast that will bear much handling when kneaded in hand it does not crumble readily but tends to work into heavy plastic, compact mass.

Simocephalus Belongs to phylum Arthropoda order cladocera family Daphnidae. The head and rostrum small, valves large and somewhat quadrate and without posterior spine, marking of transverse line over lorica.

Size limits Regulation that restrict harvest within a fishery to some specified size range of fishes.

Size of netting Size of a webbing is designated by the number of meshes in length. When a large quantity is involved the length is expressed in feet or metres of stretched webbing.

Size of twine Size (thickness) of twine is expressed by the size of yarn with which it is made and total number of yarns in the twine or number of yarns in each strand and number of such strands.

Skewness Deviation from symmetry, typically used to describe the distribution of size in a population in which the majority of individuals that are either extremely large (Positive skew) or small (negative skew).

Skiff A small boat usually 6 to 8 m long operated along with a purse seine to help in pursing the corks and to hold up the outside of the landing bag for brailing the catch.

Skinless Fish/ Skinned Fish Fish or fish fillets from which the skin has been removed.

Slaked Lime/ Calcium Hydroxide A white solid that dissolves sparingly in water to give the alkali known as lime water. Calcium hydroxide is manufactured by adding water to the oxide a process known as slaking which evolves much heat. If just sufficient water is added so that the oxide turns to a thick suspension called milk of lime is formed.

Sluice Gate A water control structure governing pond depth at a pond exit and water flow rate at a pond entrance. Screens prevent escape of stock and entry of predators and competitors. Same function as a monk.

Smoked Fish Fish cured by the action of smoke produced usually from slowly burning wood or other material (like peat) in order to partly dry the product and to give it some smoky taste.

Snag Line Wire rope which drags ahead of the trawl net to catch any obstruction in the path of the net, thus preventing damage or loss of the gear. Each end of the wire rope is attached to an otter board on its inner side.

Snood A carbon steel wire of defined length connecting the hook to the main line. This avoids direct contact.

Snout The part of the head in front of the eye.

Social risks Risks due to actions of special interest groups, such as environmentalists and conservationists.

Society of Fisheries Technologists Established in 1962. The permanent address of Society is Matsyapuri, Cochin. The Society promotes the cause for fisheries in general and fisheries technology in particular in all its aspects by providing a forum for discussion, exchange of information.

Soft Egg Disease An abnormal condition encountered in incubation fish eggs. It is also known as soft shell. The disease is identified by flaccid easily deformed character in incubating eggs. In some instances normal eggs become soft within a few days.

Soft Fibre Fibres obtained from the outer part of the stems of various dicotyledonous plants namely jute, flax, hemp, etc.

Soft Shelled A newly moulted crustacean prior to re-mineralisation of the exoskeleton. Also a pathological condition in an inter moult animal.

Soft Shelling of Prawns It is due to nutritional deficiency, pesticide contamination & poor pond water and soil condition. This disease is common in *Penaous monodon* juveniles and adults. In this, shell is thin and persistently soft for several weeks, shell surface dark, rough & wrinkled, and affected shrimps are weak. Affected individual become more susceptible to wounding, cannibalism, slow growth and eventually die.

Soil Organic Matter The organic faction of the soil that includes plant and animal residues at various stages of decomposition cells and tissues of soil organisms and substances synthesized by the soil population commonly determined as the amount of organic material contained in the soil sample passed through a 2 millimeter sieve.

Somatic Pertaining to the body.

Somatic cilia All cilia on the body surface which are outside the oral region.

Somatostain Hypothalamic releasing hormone which inhibits release of growth hormone.

Sonar A device that emits and receives sounds, used to detect underwater objects; acronym for sound navigation ranging.

Spawn It is the first feeding stage of baby fish. Its length is about 8 m.m. and weight 1.4 mg.

Spawning The natural extrusion of eggs from a female not to be confused with hatching of larvae from egg being incubated by a female.

Spawning ground Selected places to which Fishes migrate for spawning.

Spawning Ponds It is used for induced spawning of fish by simulating the natural environmental conditions.

Specific Conductance The ability of water sample to conduct an electrical current. Specific conductance values of natural water generally increases with increasing concentrations of total dissolved solids.

Specific Risk Insurance A policy which describes and pre-establishes specifically the named perils to be covered, in contrast to a policy that covers 'all risks' subject to certain exclusions.

Specific Surface Area The surface area of the material in a biofilter available for bacterial colonization usually expressed as m^2 of surface m^3 of filter volume.

Spent Fish Fish that have recently finished spawning (sometimes Shot Fish). The sexual organs are completely empty, often flaccid and blood shot. Such fish have depleted fat reserves and their flesh may have a high water.

Sperm Compound substance able to fertilize or fecundate, produced by male animals with sexual reproduction which commonly consists of the semen cells generated by the testicles and are mixed with a generally mucilaginous accompanying liquid.

Spermatophore A pocket or capsule of spermatozoa transferred to the female during mating.

Spermiation Sperm or milt production.

Sphaeromastigote The round developmental stage of a trypanosoma in which the flagellum is visible but the anterior end of the organism cannot be identified.

Spiced Cured Fish Fish cured with salt to which a mixture of spices and often sugar is added particularly herring and sprats.

Spiracle In most elasmorbranchs, the first gill-pouch between the mandibular and hyoid arches becomes modified and opens to the outside by means of a spiracle. It represents a vestigial gill pouch. The spiracle may bear rudimentary gill lamellae but these lamellae are not respiratory in function as they have only oxygenated blood supply. Such gill lamellae are called false gills or *pseudobranchs.* The spiracles open in the head region just behind the eyes. The pseudobranchs in the first pair of gill clefts of *catla catla* appear as respiratory structures in the embryonic life, but disappear later in adult. In bony fishes and lung fishes the spiracle is absent. It closes during embroynic life.

Spirodella Stem modified to flat. Size 5-10 mm, purple beneath, root fibres 2 or more, leaves absent, flower minute.

Spirogyra It is a phytoplankton belongs to family zygnemaceae. It is a silky thread like filament unbranched forming scum or mat, each cell linearly elongated with ribbon shapped chloroplast.

Splicing Joining two ends of yarn or twine or rope.

Split Fish Fish cut open from throat to vent or tail or from nape to tail, gills, guts and roe removed, head may sometimes be removed, the backbone may be left in or removed except for an inch or two at the tail for strength.

Spore The infective stage of an organism that is usually protected from the environment by one or more protective membranes.

Sporoblast A cell which develops into spores, it is produced generally by the division of a sporont.

Sporocyst A membrane bond sac which contains sporozites of Apicomplexa or the second larval stage of digneans in the gastropod intermediate host.

Sporogonic cell A cell which is produced by division of the sporoblast cell.

SporoplasmThe infections component in spores.

Sporozoite A nucleated infective stage formed by division of the sporont.

Sporting Fish A Fish that takes bait and can be caught on a hook.

Sporulation The formation of sporocysts and sporozoites within an oocyst.

Square Knot/Reef Knot/Flat Knot This knot is formed of two overhand knots with ends always falling in line with other parts.

Stagnating Standing, not flowing.

Stale Dry Gutted fish, first fermented then sundried (Burma).

Standard Pressure Meter The standard pressure meter is a solid measuring instrument for use in oceanography and limnology. In connection with vertical or horizontal plankton nets, water samplers and similar apparatus, the standard pressure meter serves for the measurement of the operational depths. Water levels and tide ranges can be measured if the instrument is used stationary.

Standard Water Sampler It is a sampler to take water samples from desired depth. The sampler remain open when lowered by rope into the water upon reaching the desired depth the messenger is let down on the rope. When it strikes the sampler the closing mechanism is released and the lids of the sampling tube close, the water sample can be drown off through the discharge cock in the lower lid for the various analyses. A thermometer (–2 to 30°C) is also fitted the sampler for reading the temperature.

Standing Crop Generally the total weights of animals of all size on a farm or in a pond.

Static Water Tank Culture tanks in which the water is changed in batches.

Steak Portion of fish cut at right angles to the backbones of the fish so that it includes a piece of the backbone may be cut from any round fish and the larger flat fish.

Stenohaline Inability to tolerate wide variations of osmatic pressure.

Sterile Not able to produce gemetes, may or may not produce gonadal hormone.

Sterilised shell fish Molluscs that have been heat treated usually by steaming or boiling to destroy non sporing bacteria (usually clams, cockles, mussels and oysters).

Steroid Hormone produced either by the gonads or adrenal glands.

Stock Group of individuals of the same species that share common production characteristics and support the same basic fisheries. Stocks are often managed as single groups of organisms even though they may be comprised of individuals from more than one population of a species.

Stock figure The number of fish that stocked in a particular pond depending on its productivity.

Stockards Solution Clearing solution used to visualise stages of embryonic development.

Strand One of the main components of braided netting yarn and rope each of which is made by twisting together filaments of wire or fibre.

Stratification The division of a water body into roughly horizontal layers of different temperature, salinity, or oxygen content.

Stratification, thermal The layering of the water column of lakes and the ocean in most system due to temperature and density, differntials between the surface and deeper waters. Warm, less dense waters near the surface mix due to wave generated currents; but deeper waters do not mix upward because they are significantly colder and dense than water at the surface

Stress The sum of the physiological response by which an animal tries to maintain or reestablish a normal metabolism in the face of a physical or chemical force. In Aquaculture it means fish disease situation i.e. when the normal functioning is significantly reduced and finally manifest in death.

Stripping The process of pressing out ripe ova or milt of fishes for artificial fecundation.

Sub-Sand Filter Water intake point for a hatchery located beneath a sandy beach in order to filtered/pretreated water supply. If gravity fed then known as a beach well.

Suberkrub Otterboard All steel chambered midwater otterboard with vertical aspects greater than its horizontal aspects named after its inventor.

Submerged Sunk under water.

Subtidal Permanently flooded habitats below the low tide line.

Sucker An adhesive organ by which a parasite attaches itself to the body of fish.

Suction Head Refers to the vertical distance from the surface of water (including draw down if any) to centre line of the pump impeller.

Sugar Cured Fish Fish particularly herring preserved with a mixture of salt and sugar.

Sun Dried Fish Fish dried by exposure to sun and wind.

Sun Hemp It is a fibre obtained from the plant *Crotalaria juncea*. Cultivated in India, Ceylon and Pakistan.

Super saturation Over abundance.

Surface Drift The method of operating of the gill net when it is allowed to drift at the surface layers of the water in such a way that the head rope of the net is exactly in line with the surface of water.

Surface feeder A fish that generally feeds in the surface layers of water.

Surface Set The method of operation of gill net when fixed for operation at the surface layers.

Surfactant A surface active agent such as soap or a detergent.

Surimi Mechanically de-boned washed and stabilized white fish flesh which is flavoured and extruded to form analogue products such as 'shrimp tails' and 'crab sticks.'

Suspended Sediment Load Particles small enough to be carried in suspension by moving water.

Suspended Solid Small particles of solids distributed through water.

Sustainable yield Manageing the harvest of a fishery resource to gain the optimum benefit to society. Differs from MSY in that sociological and economic factors as well as the biological condition and production potential of the fish stock are considered when making management decisions.

Swim Bladder Bladder containing gas lying beneath the back bone of some bony fish, marketed fresh, frozen or dried, for manufacture of isinglass and fish glue also called air bladder, Fish bladder, Fish Sound. Swim bladders when dried are also sometimes used for food, also salted.

Symbiosis Joint life of beings of different species for mutual usefulness, for instance, the presence of micro-algae in animal bodies. The union of plants and fungi to form lichens or joint life of coral fishes with certain anemones. The aim or usefulness of a symbiosis for both partners is not always clearly recognizable.

T

Table fish Edible fish of appreciable size.

Tail cloth The open trough like cloth at the hind end of the fry catching net in which fry accumulate.

Tail water A section of river downstream from a dam in which habitat conditions are modified from the natural state due to the presence of the upstream dam and reservoir.

Tangle Net Long line net suspended vertically in the water to trap fish by entangling them, also used to catch spiny lobsters and marine shrimp.

Tank One group of lakes or ponds formed by mud dams built across the valley of small streams in India. They fill during the raining season and in the Deccan plateau, provide the sole source of water.

Tannin One group of astringent aromatic acidic compounds found in the bark, root, wood, foliage or fruit of various tree and other plants. Tannin is used to convert hide substance (collagen) into leather.

Tapered Aeration Provision of more air at the entrance to an aeration tank and less near its outlet. The oxygen demand of the mixed liquor is proportional to the concentration of the unreacted sewage. Thus if all the sewage enters by the main inlet to the aeration tanks, the mixed liquor needs more air there than near the outlet. This can be achieved by having more diffusers or cone aerators of large capacity at the beginning of the tank.

Targeted Catch Same as directed catch.

Taste buds They are flask shaped multicellular structure and distributed in the epithelium of the gill arch region.

There are two distinct type of cells, sensory and supporting. They help the fish to sense the nature of food particles contained in the water flowing through the gill during ventilation.

Taxonomy The systematic classification of living things into kingdoms phyla, classes, orders, families, genera, and species.

Tea seed Cake Fish poision contains 10-15% active ingredient, saponin and is the residue after oil extraction from the seeds of camellia.

Technological Risks Risks associated with lack of adequate technology such as hatchery propagation or lack of technical information and expertise.

Technology The state of art with respect to production. It encompasses all of the known methods of production.

TED (Turtle excluding device) Mechanism placed in shrimp trawls that reduces the incidental capture of sea turtles.

Teeth in fishes A few species of fishes are tooth-less. Fishes have true teeth which are mostly of the polyphydont, aerodont and homodont variety. However, many teleosts and some elasmobranchs have heterodont teeth. The number of teeth is highly variable. It may range from a few flattened plates upto several thousand. In many species teeth may be located in various parts of the mouth or pharynx.

Teleostomi This group includes a large number of bony fishes in which the primary upper and lower jaws are supplemented by the addition of membrane bones which form the secondary jaws. Hence, they are called teleostomi or perfect-mouthed fishes.

Telson The central hind most segment of the crustacean abdomen which together with the two leaf like uropods on either side, make up the tail fan.

Temperate Zone The climatic region in the mid latitudes characterised by cold or winters and warm or hot summers.

Temperature Gradient The rates of change of temperature with distance in any given direction at any point. The average rate of · change of temperature between two points is the difference between their temperature divided by the distance between them. The vertical temperature gradient or lapse rate is the rate of change of the temperature of the air with elevation.

Territorial Water The zone of water along a coast that a country considers to be within its jurisdiction.

Tetraploid Polyploid organism or cell with four times the haploid number of chromosomes.

Textural Class The classification of soil material according to the proportion of sand, silt, and clay. The principal textural classes in soil in increasing order of the amount of silt and clay are as follows, sand, loamy sand, sandy loam, loam, silt loam, silt, sandy clay loam, clay loam, silty clay loam, sandy clay, silty clay, and clay. These class names are modified to indicate the size of the sand fraction or the presence of gravel, cobbles and stones. For example terms such as loamy fine sand, very fine sandy loam, gravelly loam, stony clay and cobbly loam are used on detailed soil maps. These terms apply only to individual soil horizons or to the surface layer of a soil type.

Texture Size and shape and arrangement of particles making up a surface deposit, rock or soil. The degree of sorting and rounding of the component particles of sediments is referred to as their textural maturity.

Thaw To melt, when ice or snow melts, it is allowed to melt

Thecodont In fishes such as garpike, barracula and haddock, teeth are fixed in deep sockets over the jaw bones. Such a condition is called thecodonty.

Theoretical Oxygen Demand The oxygen demand that would be calculated from the chemical equation if all parts of the sample were completely oxidised.

Thermal Stratification The existence within a water body of a succession of well differentiated layers, each with a different temperature, lying at various depths, the coldest at the bottom. Inverse stratification occurs beneath ice.

Thermocline /Temperature Discontinuity Layer It is a horizontal layer with a limited thickness at which temperature vertically changes abruptly. Since the differing temperatures makes the water above and below the thermocline different in their density. They do not mix easily with each other. Thus the thermocline separates vertically two water masses.

Thermophile An organism that thrives at temperature above 40°C.

Thinning Removing part of the fish stock from a pond.

Threshold The value at which a stimulus just produces a sensation, or comes just within the limits of perception, absolutely the lowest possible limit of stimulation capable of producing sensation.

Threshold Temperature The temperature at which the growth begins.

Thyroid Gland It is generally present in the pharyngeal region of all vertebrates. The gland is normally present in close association with the ventral aorta and branchial arteries in the teleosts but some times unusual positions have been reported. The function is similar like other vertebrates i.e. biosynthesis of thyroid hormone (Thyroxine). It is responsible for metabolic activity and growth. It is also responsible for the carbohydrate and protein metabolism and calcium metabolism.

Thyroxine A hormone secreted by the thyroid gland, its main function is to increase the rate of cell metabolism.

Tidal amplitude A measure of the difference between low and high tide lines in a specific region.

Tidal rip When rising or falling tides flow unidirectly through some constricted space such as the area between a submerged barrier beach and the shore line or a narrow channel connecting a large estuary to the coastline.

Tongue in fishes In fishes the tongue is merely a fleshy fold which develops from the floor of the mouth. It is devoid of muscles, however, sensory receptors may be present. The tougue is non-protrnsible. Mucous glands keep the mouthcavity slippery, and enzymes are lacking.

Topography A detailed description of the natural and artificial surface features of an area.

Total Coliform Count A bacterial count of all the different coliform bacteria present in a sample as opposed to that of one species e.g. *Escherichia coli.*

Total Dissolved Solid It is measured by evaporation at 103°–105°C of a filtered sample of water and weighing the residue, but they can also be estimated quickly from the electrical conductivity of the water, continuous recorders of conductivity immediately indicate any change in conductivity that might be a sign of new pollution TDS is the rough indicator of edaphic conditions, which affects the productivity of the water bodies.

Total Dynamic Head It is the sum of the suction head, discharge head, hydraulic head losses and the velocity head.

Total length The greatest straight line distance between the tip of the snout and the tip of the tail fin.

Total Non-Volatile Solids The difference between the solids and total volatile solids. The difference represents the organic substances.

Total Organic Carbon A measure of the amount of carbon in a sample originating from organic matter.

Total Oxygen Demand It can be measured from the depletion of oxygen in a nitrogen oxygen carrier gas in a platinum catalysed combustion chamber.

Total Solids The residue left upon evaporation of a raw water sample. This residue includes dissolved gases, carbondioxide, volatilised from the biocarbonates and all suspended matter.

Total Volatile Solids The loss of weight occurring ignition at 550°C of the total solid residue. This weight loss represents primarily organic matter.

Toxic Waste Poisonous waste. That is illegal to transport or dip it without notifying the local authority in the area where it is, the local authority in the area where it will be dipped and the water authorities in both areas.

Toxicity Poisonous property.

Tran Oil Oil obtained from fish and aquatic animals by allowing them to decompose. The term "tran" in various languages also applies to the uncleaned fish fat or fish oil.

Trapa It is commonly known as water chestnut, plants floating usually rooting in the mid, with leaves clustered toward the tip of the stem, leaves with rather 4 sided blades and long petioles which have an inflated spongy region fruit nut like 4 horned, edible.

Trapnet Passive capture gear with an intrapment area consisting of a series of funnels by fishing vessels.

Trash Fish/Industrial Fish Fish or quantities of mixed species incidently caught and unwanted for human consumption.

Treflan—R It is a strong fungicide originally developed as a herbicide.

Tributary A stream or other body of water that contributes its water to a larger body of water.

Trichophrya This is a small parasite which resembles pin cushions and are found on the gills, delicate tentacles, extend out from the true parasites but their presence is irritating to the gills and they may become abundant enough to interfere with respiration.

Trickling Filter / Biological Filter An aerobic process used in secondary treatment plants for the processing of sewage. Trickling filters consists of beds of coarse material over which the sewage is sprinkled at a uniform rate, they are generally not less than two meters deep and circular in plan.

Tricodina A smaller circular parasites with a conspicuous teethed disc within their body, with a dart erractically about over surface of body gills and fins. Cilia around margin may or may not be obvious, cat fish, gold fish, common carp, channa sp. served as the host for the parasite.

Trimming Removal of inedible parts or any thing which could spoil the appearance.

Troncon Portion of fish cut perpendicularly to the back bone, should have a thickness of at least the width.

Trophic depression The decline in production seen in reservoirs as they age. This decline is largely due to a reduction in leaching a soluble nutrients from soils to the water shed into the reservoir and from the loss of nutrients within the reservoir due to its high flushing rate.

Trophozoite The feeding vegetative and non dividing stage in parasitic protozoa it is also called a trophont.

Tropic upsurge Initial pulse of high levels of productivity in a newly filled reservoir due to large part to an influx of nutrients leaching from newly bared and flooded soils and from the decomposition of substantial amounts of organic debris left when the reservoir basin was deforested.

Tuna Ham Smoked fish sausage, the encased mixture of tuna meat, salt, sugar, starch and spices is smoked for 12 hours before vaccum packaging, commonly packed in cylindrical or square plastic casings.

Turbidity State of water not being clear or translucent due to suspended fine particles in it.

Turn-over The movement of surface waters to the substrate and deeper water towards the surface in standing bodies of water such as lakes and the ocean. In many system turnover occurs in the spring and fall due to differences in density between the surface and deeper waters and due to vertical currents created by wave action.

Two story Lakes or reservoirs Systems that have substantial volume of deep water below the hypolimnion that remain suitable habitat for cold water fish species through out the year, as well as surface water habitats that support warm water and cold water species.

Typha This is an marginal aquatic plant, it may appear in almost any wet place and often the first invaders in a newly excavated pool. The underground stems spread extensively, stem slender, leaves flat and somewhat round in the back, upper part projects out of the water to lengths 10-12 feet, rounded and joined.

U

Ulcer The excavation of the surface of an organ or tissue and is produced by sloughing of necrotic inflammary tissue.

Ulceration A Pathological condition which is usually associated with the skin or mucosal surfaces it involves lesions with erosion of surface epithelia and inflammation with infiltration of leucocytes.

Ulothrix It is a phytoplankton belongs to order ulotrichales and family ulotrichaceae. It is filamentous unbranched cells, brick shaped with single girdle shaped chloroplast.

Ultimate Oxygen Demand, UOD Either the theoretical oxygen demand or the ultimate biochemical oxygen demand, the maximum value of the BOD test ordinarily reached after 25-30 days incubation. The UOD is the sum of the ultimate carbonaceous BOD and the ultimate nitrogenous BOD.

Ultra Violet Radiation/UV Electro magnetic radiation in the wave length range about 4×10^{-7} to 5×10^{-7} meters, placing it between visible light and X-ray in spectrum. The UV spectrum is divided further by wavelengths into A, B, and C bands and certain wave length that affects cells are known in USA as DUV (Damaging UV). Much of the UV radiation from the Sun is absorbed in the ozone layer of the earth's atmosphere. UV radiation affects photographic films and the skins of some animals, including man it acts on ergosterol to produce vitamin D. UV is non-ionising radiation with less penetrative power in a column of water than visible light, but with in cell it causes chromosome breaks. In nature, most organisms are well protected against harmful UV. In man the radiation is absorbed by melanin in the skin causing the melanin to darken. UV can be produced artificially by mercury vapour lamps.

Umbrella Net Umbrella shaped lift net.

Under Development Any condition in which full industrialisation and the full use of the resource base not occurring.

Under Flow In sewage treatment the flow from the bottom of the tank e.g. sludge from a secondary sedimentation tank or gravity sludge thickner. In mineral dressing it may be dense slurry that flows out of the bottom of a gravity thickner or the undersize from a wet classifier.

Under stocking Stocking below the capacity of the pond.

Unsaturated Fatty Acid Fatty acids that contain less than the maximum proportion of hydrogen relative to the amount of carbon present. These hydrogen deficient fatty acids are very reactive. Their occurence in fats has pronounced depressing effects on the melting point and storage life of the fat.

Upland Catchment Water Water from land that is more than 100-200 meters above sea level is likely to be unpolluted by sewage, because most large cities are near sea level. It is therefore impossible for much sewage to flow into upland streams. Surface upland water especially from moorland, is likely to be bacteriologically pure enough for drinking, but it may contain organic acids if the ground is peaty and may also soft and plumbo-solvent.

Upwelling The upward flow of water that occurs when ocean currents strike a continental coastline and deflect laterally. This defections causes water to rise upward from the ocean depths to the surface.

Ureotelic Term applied to those animals whose main nitrogenous excretory product is urea. Fish, amphibians and man are the ureotelic. During their embryonic development dissolved urea diffuses away into the surrounding water or is removed in the maternal blood stream.

Uric Acid A complex nitrogenous compound, an almost insoluble breakdown product of proteins and nucleic acids, excreted in large amounts by ureotelic animals and in small amount by some other animals.

Utricularia Small herbs growing in the water or in damp places with bladders to catch small animal life except in some terrestrial species. Flowers yellow or purple, two lipped the upper lip erect, the lower lip with a spur and usually with a prominent palate. Stem soft, cylindrical and branched roots hairy, leaves usually filiform or finally dissected.

UV Sterilization Ultra violet irradiation of water to kill or inhibit bacterial development. Only likely to be effective when the water is free of particulate material which the light can not penetrate.

V

Vaccination Inoculation of the host with a vaccine (Usually made from dead or attenuated virus or bacteria) which provokes an immune response in order to protect the host from subsequent disease due to the organism in question.

Vaccine A preparation of non virulant disease organisms or immunogens which still retain the capacity to stimulate the production of antibodies or resistance.

Vallisneria It is a submerged type aquatic plant, stem berried in the mud, sending up tufts of ribbon like leaves in the water in clusters. Staminate flowers in a spathe which breaks and allows them to come to the surface where they float.

Value Added Processing or presenting crustacean, fish flesh in a more attractive way to increase its market value.

Variable Cost Cost that vary with the level of output, such as payments for inputs. Average variable cost refers to variable cost per unit of output.

Variation Differences which exist between individuals of the same species at corresponding stages in the life cycle, variation is a reflection of the genetic constitution of the individuals and the environmental influences brought to bear upon them.

Variety A group of plants or animals which differ distinctly from other within the same sub-species.

Ventral Pertaining to the lower surface of the body.

Ventral fin The paired fins situated below and behind the pectoral fins, also called the pelvic fins.

Venture Capital The portion of capital raised by the sale of equity.

Vertebrate An organism with a back bone of vertebrae. The vertebrate includes fish, amphibians, reptiles, birds, and mammals.

Vertical Incubator Container for incubating eggs where water enters at the bottom and flows upwards through the eggs.

Vertical Scale Disease /Pine Cone Disease This disease mainly infects common carp, grass carp, silver carp etc. The skin of the diseased fish appears rough some scales (especially on the posterior part of the body) are stretched out resembling pine cones. The scale capsule contains a semiopaque liquid that makes the scales vertical. If slight pressure is applied to the scale the liquid will exude from the scale base and the scale will immediately drop off. Other symptoms include congestion on the finbases, mild bleeding and inflammation on the skin, reddish ulceration on the desquomated area, exophthalmos and abdominal distension. As the disease develops fish swim slowly, show dyspnea and the abdomen turns upward. The fish will die in 2-3 days later, the causative agent is *Pseudomonas punctata.*

Vertical Transmission Infection transmitted from parent fish to progeny via infected eggs.

Vinegar Cured Fish Fish preserved in medium containing vinegar and salt with or without spices.

Viral Hemorrhagic Septicemia (VHS) It is caused by virus. The principal symptoms are kidney swelling, pop eye disease, abdominal dropsy, reduced appetite, erratic spiral swimming, internal multiple hemorrhages in skeletal muscle ocular connective tissues and viscera. Kidney usually hypermic and may be swollen. Liver may be hypermic grey. Digestive tract usually empty. Fish fail to maintain balance, swim in circles and make frenzid leaps interpreted as motor disorders.

Virgin Soil A soil that has not been significantly disturbed from its natural environment.

Virgin water Water not previously utilized for systematic fish culture.

Viscocity The property of being thick and not flowing quickly.

Vitamins One group of chemicals constituents of foods in their natural state of which very small quantity are essential for the normal nutrition of animals and plants. They are accessory food factors that must be supplied for the nutrition of an organism and that exert an important action in the control and coordination of physiological functions by virtue of specific molecular structure.

Vivier truck Vehicular transport equipped with facilities for the live transportation of adult crustaceans.

Viviparous fishes Species in which embroys are incubated with in the reproductive tract of the female, and are nourished through a capillary attachment female and embryo.

Volvox It is a phytoplankton belongs to family volvocaceae. It is rounded, colonial form consisting of 500-60000 cells. Marginal cells pyriform while centre cells rounded in shape, several flagella coming outside, gelatinous mass.

W

Waste Any substance solid or liquid or gaseous for which no use can be found by the organism or system that produces it and for which a method of disposal must be devised.

Waste Water Treatment Plant A series of tanks screens filters and other processes by which pollutants are removed from water.

Water Analysis The determination of the physical, chemical and biological characteristics of the water. Such analysis usually involve four kinds of examination, physical, chemical, bacterial and microscopic.

Water Balance It is assumed that in the long term a water budget will balance and the water balance lists the incoming and outgoing water.

Water Budget A budget of incoming and outgoing water from a region including rain fall, evaporation, run off, seepage and with perhaps, special attention to ablation of snow, evapotranspiration from vegetation, dew etc.

Water Colour The colour of water always coming from vegetation or other organic matter comes in contact with water. Water colour can be measured either by a colorimeter or by comparing it with standard solutions of known colour. Good distilled water has a colour of zero, as many well waters.

Water fleas Micro Crustaceans largely used as food by baby fish.

Water Quality The physical, chemical, and biological characteristics of water in respect to its suitability for a particular purpose. The same water may be of good quality for one purpose or use and bad for another.

Water Shed The total area of a given point on a stream that contributes water to the flow at that point. Synonyms are drainage basin or catchment basins.

Water Table The upper surface of the saturated rocks in which all voids are filled with water, water table are rarely flat and tend to follow the relief of the land surface but with undulations smoothed out. They are generally deeper on the crest of a hill than in a valley bottom. The water table intersects the ground surface, may emerge above ground in the form of springs, marshes or lakes.

Water Thermometers It is used for temperature measurements to be taken in deep running or stagnent water as well as in wells and industrial tanks. A scoop made of brass with perforated top, 70 mm high and 32mm in diameter, which is filled with water during measurement is mounted on a brass armour size 200 × 20 mm. The thermometer to be kept at desired level until it has taken up the temperature prevailing there. Then it has to be pulled up quickly, where by the temperature absorbed in the depth is retained by the thermometers in consequences of the thermal lag resulting from the strong thermal capacity of the scoop. The window through which the reading are taken may be closed by another revolving metal case so as to protect the instrument during measurement.

Water Turbidity Relatively inert and finely divided material are easily suspended in water, cutting down light transmittance enough to inhibit photosynthesis of both micro and macro vegetation.

Water Yield The run off from the drainage basin, including ground water outflow that appears in the stream plus ground water outflow that bypasses the gauging station and leaves the basin underground. Water yield is the precipitation minus the evapotranspiration. It is also known as the water crop or run out.

Weed Any wild plant or animals with little or no commercial value that develops on cultivated ground and excludes or damages the desired crop.

Weed fish Minnows and other fishes which serves as food of other selected fishes in Pisciculture.

Weir A permanent passive capture gear setup in which fishes wonder and donot find their way out.

Wet bundh A perennial irrigations pond used for carp breeding.

Wet Land An area covered permanently, occasionally, or periodically by fresh or salt water upto a depth of 6 meter.

Whale Oil It is obtained from the bladder and other parts of the body including bones, of whales and similar marine animals may be used in natural state for industrial purposes, but mainly used after hydrogenation and refining for manufacture of edible fats.

Whirling Disease It is caused by Myxosoma cerebvalis. The symptoms are rapid tail chasing type of whirling, is often seen when the fish is frightened or trying to feed. The whirling symptom is associated with lesions and disintegration of the cartilaginous skeleton support of the organs of equilibrium caused by invasion of the parasite protozoan.

Whirlpool A circular eddy in any body of water.

White Fish Fish in which the main reserves of fat are in the liver. The term is common in U.K.

White Fish Meal Fish Meal made from surplus non fatty or from processing waste from such fish.

White Wings Dried salted split cod that has the black lining removed from the belly walls of the split fish.

Wild Type The most frequently encountered phenotype in natural breeding populations.

Wind Dried Fish Fish dried naturally in the wind.

Wing Part of the body of some muscled flat fish e.g. the edible part of skates in France also refer to the belly walls of salted cod.

Wolffia It is a floating aquatic plant, stemless and rootless, star shaped green leaves clumped together. It consists of almost microscopic meal like bodies without roots. Flowers small. It is the smallest flowering plant.

World Bank International Bank for reconstruction and development.

Worm Cataract Whitish eye due to the white dot i.e. little worms situated between lens and lens capsule. It is caused by *Hemistomum spathaceum.*

X

Xanthophyta (Yellow green algae) A group of mainly fresh water and terrestrial algae which contain carotenoid pigments as well as chlorophyll. Most are unicellular and non motile, some are colonial or filamentous. A few are colourless and saprophytic or ingest food particles like protozoa.

XBT Expendable bathythermograph, an instrument which is lowered from the ship to measure temperature changes with depth and is not retrieved.

Xenoma A symbiotic complex formed by a hypertrophic host cell and an intracellular parasite which proliferates in its Cytoplasm.

Y

Yearlings A one year old fish.

Yeast Many species of unicellular fungi, most of which belongs to the ascomycetes and reproduce by budding. The genus Saccharomyces is used in wine making because in low oxygen concentrations it produces zymase, an enzyme system which breaks down sugars into alcohol and carbondioxide, Saccharomyces also used in baking. Some yeasts are used as a source of protein and vitamins of the B-group.

Yellow Cure Refer Amarelo Cure.

Yellow Fish Any white fish, split or filleted cold smoked. The term is commonly used in UK.

Yield Harvested portion, a population sustainable yield is the portion of a population that can be removed by fishing and replaced by the following year through reproduction and body growth.

Yolk Sac An embryonic memberane which encloses and absorb the yolk in reptiles, fish and many birds.

Z

Zeolite Softener An ion exhange process used to soften water, consisting of replacing the calcium and magnesium ions with the sodium ion. Sodium chloride is used for the regeneration. When operated on the hydrogen cycle, calcium and magnesium are replaced with the hydrogen and a mineral acid (usually sulphuric acid) is used for regeneration.

Zonal Index A measure of the strength of the circulation in a specified large area of the globe, often expressed in the form of a pressure differences.

Zonal Soil Those soil that have mature profiles developed under specific conditions of climate and vegetation.

Zone of Aeration The zone above the water table. Water in the zone of aeration is under atmospheric pressure and will not flow into the well.

Zone of Saturation The space below the water tables in which all the interstices are filled with water. Water in the zone of saturation is called ground water.

Zoosis Any disease caused by a parasitic animal.

Zooplankton Microscopic animals that are suspended in the water column. Although many are capable of locomotion, they will drift horizontally in the water column with prevailing currents.

Zoonosis A disease shared in nature by man and other Vertebrates.

Scientific name of some common fishes

Common name	Scientific name	Countries
Alfonsino	Beryx splendens	Cosmopolitan
Alka Mackerel	Pleurogrammus azonus	Japan
Allis shad	Alosa alosa	N. Atlantic
American eel	Anguilla rostrata	Atlantic
American Plaice	Hippoglossoides plalessoides	N. Atlantic
American shad	Alosa sapidissima	Atlantic
Angel	Frozen turtle	U.K.
Angel fish	Squatina squatina	U.K.
	Brama brama	U.K.
	Holacanthus & Pomacanthus spp	America
Angler fish	Lophius spp	Cosmopolitan
Lesser Silver smelt	Argentina sphyraena	Atlantic
Great Silver smelt	Argentina silus	N. Atlantic
Deep Sea smelt	Argentina semifasciata	Pacific
Silver side	Argentina elongata	Australia Newzealand
Arrow tooth flounder	Atheresthes stomias	Pacific
Arrow tooth halibut	Atheresthes evermanni	Pacific
Axillary Bream	Pagellus acarne	Atlantic
Ayu sweet-fish	Plecoglossus altivelis	Japan Korea, Taiwan China
Barded Barb	Barbus partipentazone	
Barrnacle	Balanus spp	
Beaked whale	Ziphias mesoplodon	
White whale	Delphinapterus leucas	Artic
Big eye	Priacanthidae member	Atlantic
Big eye Tuna	Thunnus obesus	Cosmopolitan
Big skate	Raja binoculata	Pacific
Bill fish	Istiophoridae member	Cosmoplitan
Black Bass	Micropterus spp	N. America
Black drum	Pogonias cromis	Atlantic
Black mouthed Dog fish	Galeus melastomus	Atlantic
Black Ruby barb	Barbus nigrofasciatus	
Black shark	Members of Squalidae family	Cosmopolitan

Common name	Scientific name	Countries
Bleak	Alburnus alburnus	Europe
Blind fish	Members of family Amblyopsidae	
Blind cave fish	Anoptichthys Jordani	
Blue crab	Callinectes sapidus	Atlantic
Bluefin Tuna	Thunnus thynnus	Cosmopolilitan
Blue mussel	Mytilus edulis	N. Atlantic
Boar fish	Members of Pentacerotidae family	Pacific
Bombay duck	Harpodon nehereus	India
Bonga	Ethmalosa fimbriata	West Africa
Bomito	Sarda spp	Cosmopoitan
Bora fish	Pisodonophis spp.	
Bottle nosed Dolphin	Tursiops truncatus	Cosmopolitan
Bottle nosed whale	Hyperoodon rostratus	N. Atlantic
Bow fin	Amia calva	USA
Bream	Abramis spp.	Europe
Brill/Brett/Brit/Kito/Pearl	Scophthalmus rhombus	North sea to mediterranean
Bronze whale	Carcharhinus brachyurus	Australia Newzealand
Brown Algae	Laminaria spp	
Brown cat shark	Apristurus brunneus	Pacific N. America
Brown shrimp	Crangon crangon	Europe/Africa
Painted River Prawn	Macrobrachium carcinus	America
Yellow leg shrimp	Penaeus californiensis	America
Witch Prawn	Penaeus canaliculatus	Indian Ocean
Buddha's Ear	Iridea laminaroides (Sea weed)	Japan
Bullet Tuna	Auxis rochee	Cosmopoilitan
Bull frog	Rana catesbeina	Japan
Bull shark	Carcharhinus leucas	N. America
Butter fish	Stromateidae member	Cosmopolitan
Butter fly fish	member of Chaetodontidae	Cosmopolitan
Mullet	Mugil spp	
Pearl spot	Etroplus spp	
Thread fin	Polynemus spp Polydactylus spp	

Common name	Scientific name	Countries
Milk fish	Chanos spp	
Chum	Oncorhynchus keta	Pacific/N. America
Coal fish	Pollachius virens	Cosmopolitan
Cobbler	Cnidoglanis macrocephalus	Australia
Coho	Oncorhynchus kisutch	N. America
Comber	Serranus cabrilla	Atlantic Red Sea
Common Dolphin	Delphinus delphis delphis	Cosmopolitan
Craw fish	Palinurus panulirus	
Cream fish	Parika scaber	New Zealand
Crucian carp	Carassius carassius	
Cuttle fish	Sepia spp	Europe/Asia
Chub mackerel	Scomber Japonicus	Cosmopolitan
Dara	Thread fin	Cosmopolitan
Devil fish	Mobula mobular	India
Dolphin fish	Coryphae midae	Atlantic Mediterranean
Pompano Dolphin	Coryphaena equisetis	Tropical & sub tropical
Double lined mackerel	Grammatorcynus bicarinatus	Atlantic America
Drum	member of Sciaenidae	Indo-Pacific
Dulse	Rhodymenia palmata (Red algae)	Cosmopolitan Atlantic
Dusky Dolphin	Lagenorhynchus obscurus	S. Atlantic
Dwarf Gourami	Colisa lalia	
Eel	member of Anguillidae	Cosmopolitan
Elephant fish	Callorhynchus callorhynchus	S. America
European eel	Anguilla anguilla	Arctic mediterranean
Feather Backs	Member of family Netopteridae	
Flat fish	Member of Heterosomata	
Flat head	member of Percophidae	Cosmopolitan
Flat head skate	Raja rosispinis	Pacific
Fluke	Paralichthys spp	Atlantic
Flying barb	Member of Genus Esomus	
Flying fish	member of Exocoetidae	Cosmopolitan
Flying squid	Todarodes sagittatus	Atlantic/Europe
Freshwater eels	Member of family Angullidae	
Frost fish	Lepidopus caudatus	Cosmopolitan
Gaint Gourami	Colisa fasciata	
Gar fish	Belone belone	N. Atlantic Europe

Common name	Scientific name	Countries
Gem fish	Rexea solandri	Australia / New Zealand
Gill head bream	Sparus aurita	E. Atlantic
Gilt sardine	Sardinella aurita	Mediterranean
Gizzard shad	Dorosoma copedianum	Atlantic
Goat fish	Member of Mullidae	Cosmopolitan
Gold fish	Carassius auratus	Cosmopolitan
Gold line	Sarpa salpa	Atlantic
Gourami	Osphronemus goramy	India/Malaya
Greater weever	Trachinus draco	Atlantic mediterranean
Green laver	Enteromorpha linza	Japan
Guitar fish	Member of Rhinobatidae	Cosmopolitan
Gummy shark	Mustelus spp	
Haddock	Melanogrammus oeglefinus	N. Atlantic
Half beak	Hemiramphidae	Atlantic/Pacific
Half moon	Member of Kyphosidae	Pacific
Hammer headed shark	Member of Sphyrnidae	Cosmopolitan
Hard clam	Mercenaria mercenaria	Atlantic America / Japan
Harvest fish	Peprilus spp	Atlantic / N. America
Hen fish	Cyclopterus lumpus	
Houting	Coregonus oxythynchus	Atlantic North sea
Humantin	Oxynotus centrina	Atlantic Europe
Hump back whale	Megaptera novaenglia	Cosmopolitan
Ide	Leucisus idus	Europe
Indian mackerel	Rastrelliger spp.	Indian & Pacific ocean
Indian porpoise	Neomeris phocaenoides	Indo-Pacific
Irish moss	Chondrus crispus (Red algae)	
Jack Mackerel/Horse mackerel	Trachurus & Decapterus spp.	
Japanese eel	Anguilla Japonica	Japan
Japanese pilchard	Sardinops melanosticta	Japan
Japan Sea Bass	Lateolabrax japonicus	Japan
Jelly fish	Rhopilema spp.	Japan
Jew fish	Epinephelus itajara	Atlantic N. America
Joey	Small mackerel	Cosmopolitan
Jhon dory	Zedae	Cosmopolitan
Kilkar	Clupeonella delicatula	
Killer whale	Orcinus orca	Atlantic

Common name	Scientific name	Countries
Killi fish	Member of Cyprinodontidae	Pacific
King crab	Paralithodes camchatica	
King fish	Scomberomrus cavalia	Atlantic
King klip	Genypterus capensis	South Africa
King Mackerel	Scomeromorus spp.	
Kombu	Laminaria spp. (Sea weed)	Japan
Lady fish	Elops saurus	Atlantic N. America
Lemon sole	Microstomus kit	N.E. Atlantic / North Sea
Limpet	Patella caerulea	Atlantic, Europe
Lined sole	Archirus lineatus	Atlantic, N. America
Little skate	Raja erinacea	Atlantic N. America
Little tunny	Euthynnus alletteratus	Atlantic / Mediterranean
Lizard fish	Sauridae undosquanis	Cosmopolitan
Lobster	Homarus spp.	Atlantic
Long nose skate	Raja oxyrhinchus	Atlantic
Lump fish	Cyclopterus lumpus	N. Atlantic/Arctic
Mackerel	Scomber spp.	Cosmopolitan
Mackerel shark	Lamnidae member	Cosmopolitan
Marlin	Makaira spp.	Cosmopolitan
Meagre	Argyrosomus regius	E. Atlantic
Milk fish	Chanos chanos	Cosmopolitan
Minke whale	Balaenoptera acutorostrata	Cosmopolitan
Moki	Latridopsis ciliaris	Indian & Pacific
Mola	Member of Molidae	Cosmopolitan
Monk fish	Lophius piscatorius	
Moon fish	Vomer spp.	N. America
Mud fish	Amia calva	
Mullet	Member of Mugilidae	
Needle fish	Member of Belonidae	Cosmopolitan
Nurse shark	Ginglymostoma spp.	Atlantic
Oar fish	Regalecus glense	Atlantic
Oil sardine	Sardinella longiceps	India Pakistan Philippines
Orange perch	Lepidoperca spp.	New Zealand
Paddle fish	Member of Polydontidae	USA
Painted Ray	Raja microocellata	Atlantic-Europe
Parrot fish	Member of Scaridae	Cosmopolitan
Pig fish	Orthopristis chrysoptera	Atlantic
Pigmy whale	Carperea marginata	Southern Seas

Common name	Scientific name	Countries
Pike Pearch	Lucioperca & stizostedion spp.	Europe/N. America
Pike	Esocidae	Europe/N. America
Pilot whale	Globicephala spp.	Cosmopolitan
Pin fish	Lagodon rhomboides	Atlantic/N. America
Pink salmon	Oncorhynchus gorbuscha	Pacific
Pipi	Plebidonax deltoides	Australia
Black pomfret	Stromateus niger	Indian ocean
White pomfret	Stromateus cinercus	Indian ocean
Pond smelt	Hypomesus olidus	N. America/Japan
Giant tigar prawn	Penaeus monodon	Indo-Pacific
Indian prawn	Penaeus indicus	Indo-Pacific
Banana prawn	Penaeus merguiensis	Indo-Pacific
Puffer	Tetraodontidae member	Cosmopolitan
Queen fish	Seriphus politus	Pacific N. America
Rabbit fish	Chimaera monstrosa	N. E. Atlantic
Rat fish	Hydrolagus colliei	Pacific N. America
Red fish	Sebastes spp.	U.K.
Ruby fish	Plagiogeneion rubiginosus	Indian ocean
Sable fish	Anopoploma fimbria	Pacific
Sail fish	Istiophorus spp.	Cosmopolitan
Sand fish	Member of Trichondontidae	Pacific
Saw fish	Member of Pristidae	Cosmopolitan
Scorpion fish	Scorpaena spp.	Mediterranean Atlantic
Sea cabbage	Laminaria spp.	
Sea cow	Dugong spp.	Indo-Pacific
Sea cucumber	Member of Holothuroidae	Cosmopolitan
Sea trout	Salmotrutta	Atlantic/N. America
Sea trumpet	Eisenia bicyclis	Japan
Sea Urchin	Echinus spp.	Cosmopolitan
Shoe Maker	Tanca tinca	
Smelt	Member of Osmeridae	Cosmopolitan
Spear fish	Tetrapturus spp.	Cosmopolitan
Sperm whale	Physeter macrocephalus	Cosmopolitan
Star fish	Member of Asteroidea	Cosmopolitan
Sword fish	Xiphias gladius	Cosmopolitan
Tench	Tinca tinca	
Garai	Channa (ophiocephalus) spp.	

Common name	Scientific name	Countries
Magur	Clarius batrachus	
Penga	Pangasius pangasius	
Tengra	Mystus spp.	
Pavda	Ompak spp.	
Boari	Wallago attu	
Koi	Anabus testudienus	
Bele	Glossogobius girius	
Chital	Notopterus chitala	
Guchia	Amphipnous guchia	
Bami	Mastocembelus spp.	
Pothia	Puntius spp.	
Chelwa	Oxygastor spp.	
Chanda	Ambasis spp.	
Mola	Amblypharyngodon mole	
Colisa	Colisa spp.	
Lebuca	Chela lebuca	
Chera	Gudusia chapra	
Esomus	Esomus danricus	
White carp	Cirrhina cirrhosa	
Pig mouth carp	Labeo kontius	
Bata	Labeo bata	
Doctor fish	Tinca tinca	
Pearl spot	Etropus surantensis	
Grey mullet	Mugil cephalus	
Thick back sole	Microchirus variegatus	
Thorn back ray	Raja clavata	Atlantic Mediterranean
Tope	Galeorhinus galeus	
Trigger fish	Member of Balistidae	
Tuna	Balistes carolinensis	
	Members of thunnidae	Cosmpolitan
	Thunnus spp.	
Tusk	Brosme brosme	
Weark fish	Cynoscion spp.	North Atlantic
		Atlantic/Pacific North &
Whelk	Buccinum undatum	South America
Whiptail	Stingray group	
White Amur	Ctenopharyngodon idella	
White Bass	Morone chrysops	N. America

Common name	Scientific name	Countries
White bream	Diplodus sargus	Atlantic Mediterranean
White croaker	Argyrosomus argentatus	
White Hake	Urophycis tenuis	Atlantic N. America
White perch	Morone americanus	Atlantic
White shark	Carcharodon carcharias	Cosmopolitan
White shrimp	Penaeus spp.	Cosmopolitan
White sided Dolphin	Lagenorhychus acutus	North Atlantic
White skate	Raja alba	Atlantic mediterranean
White tip shark	Carcharhinus longimanus	Atlantic N. America
Witch	Glyptocephalus cynoglossus	North Atlantic
Wolf fish	Anarhichas spp.	North America
Wreck fish	Polyprion spp. (Polyprion americanus)	
Yellow eye mullet	Aldrichetta forsteri	Australia/Newzealand
Yellow fin tuna	Thunnus albacares	Cosmopolitan
Yellow gurnard	Trigla lucerna	Atlantic Mediterranean
Yellow perch	Perca flavescens	Japan & China
Yellow sole	Buglossidium luteum	Atlantic Mediterranean
Yellow tail king fish	Seriola grandis	Australia
Yellow tail	Seriola spp.	Cosmopolitan
Yellow sea Bream	Taius tumifrons	Japan & China
Zanthe	Abramis vimba	Europe
Zebra fish	Member of Genus Danio	

Table 1 : Conversion of volume (U.S. Imperial, and Metric equivalents Length, Weight, Area and Flow rates

VOLUME

Litres	×	0.264	=	Us gallons
Us gallons	×	3.785	=	Litres
Imperial gallons	×	1.201	=	Us gallons
Imperial gallons	×	4.546	=	Litres
Us gallons	×	0.8327	=	Imperial gallons
Cubic meters	×	35.31	=	Cubic feet
Cubic feet	×	0.0283	=	Cubic meters
Cubic feet	×	28.316	=	Litres
Cubic inches	×	16.3871	=	Cubic centimeters
Cubic feet	×	264.2	=	Us gallons
Us gallons	×	0.00379	=	Cubic meters
Acre feet	×	1233.48	=	Cubic meters

LENGTH

Meters	×	3.281	=	Feet
Feet	×	0.3048	=	Meters
Meters	×	39.37	=	Inches
Inches	×	0.0254	=	Meters
Centimeters	×	0.3937	=	Inches
Inches	×	2.54	=	Centimeters
Millimeters	×	0.0394	=	Inches
Inches	×	25.4	=	Millimeters
Miles	×	1.6093	=	Kilometers

WEIGHT

Kilograms	×	2.2046	=	l bs.
Lbs.	×	0.4536	=	Kilograms
Tons (long)	×	1016.05	=	Kilograms
Tons (long)	×	2240	=	l bs.
Tonnes (Metric)	×	1000	=	Kilograms
Tonnes (Metric)	×	2204.6	=	l bs.
Tons (short)	×	907.185	=	Kilograms
Tons (short)	×	2000	=	1bs.
Grams	×	15.432653	=	Grains
Grains	×	0.0647989	=	Grams
Grams	×	0.0352740	=	Ounces (Us)
Ounces (Us)	×	28.349527	=	Grams
Cubic feet	×	62.414	=	l bs (at 4°C)
Cubic feet	×	62.234	=	l bs (at 25°C)
Cubic feet	×	2719000	=	l bs (at 4°C)
Acre feet	×	2711000	=	l bs (at 25°C)

AREA

Square inch	×	6.4516	=	Sq. centimeters
Sq. feet	×	0.09290	=	Sq. Meters
Acre	×	0.404686	=	Hectares
Hectares	×	10,000	=	Sq. meters
Sq. mile	×	2.5900	=	Sq. kilometers
Acre	×	43560	=	Sq. feet
Sq. feet	×	144	=	Sq. inches
Sq. yard	×	9	=	Sq. feet
Acre	×	4840	=	Sq yard

FLOW RATES

gal/min	×	0.00223	=	Cubic feet/sec.
gal/min	×	0.00442	=	Acre foot/day
gal/min	×	0.0631	=	lit/sec.
gal/min	×	5.42	=	Cubic /day
Cubic feet/sec.	×	449	=	gal/min.
Cubic feet/sec.	×	1.98	=	acre feet/day
Cubic feet/sec.	×	28.3	=	lit/sec.
Cubic feet/sec.	×	0.0283	=	Cubic meter/sec.
Cubic feet/sec.	×	2450	=	Cubic meter/day
Acre foot/day	×	226	=	gal/min.
Acre foot/day	×	0.505	=	Cubic feet/sec.
Acre foot/day	×	14.2	=	Litre/sec.
Acre foot/day	×	1230	=	Cubic metre/day
Litre/sec.	×	15.9	=	gal/min.
Litre/sec.	×	0.0353	=	Cubic feet/sec
Litre/sec.	×	0.0703	=	Acre foot/day
Litre/sec.	×	86.4	=	Cubic metre/day
Cubic meter/sec.	×	15.800	=	gal/min
Cubic meter/sec.	×	35.3	=	Cubic feet/sec.
Cubic meter/sec.	×	70.0	=	Acre feet/day
Cubic meter/sec.	×	86400	=	Cubic meter /day

Table 2 : Factors for converting conductivity of water to conductivity at 25° C

Tem.°C	*Factor*	*Temp. °C*	*Factor*	*Temp.°C*	*Factor*
10	1.36	19	1.12	28	0.95
11	1.33	20	1.10	29	0.93
12	1.30	21	1.08	30	0.92
13	1.27	22	1.06	31	0.90
14	1.24	23	1.04	32	0.89
15	1.21	24	1.02		
16	1.19	25	1.00		
17	1.16	26	0.98		
18	1.14	27	0.97		

Table 3 : Solubility of Oxygen

Water temperature (°C)	*Solubility of Oxygen (ppm)*
0	14.16
1	13.77
2	13.40
3	13.05
4	12.70
5	12.37
6	12.06
7	11.76
8	11.47
9	11.19
10	10.92
11	10.67
12	10.43
13	10.20
14	9.98
15	9.76
16	9.56
17	9.37
18	9.18
19	9.01
20	8.84
21	8.68
22	8.53
23	8.38
24	8.25
25	8.11
26	7.97
27	7.86
28	7.75
29	7.64
30	7.53
31	7.42
32	7.32

Table 4 : Percentage unionized Hydrogen Sulphide in water at different pH and Temperature

Temperature									
pH	*16*	*18*	*20*	*22*	*24*	*26*	*28*	*30*	*32*
5.0	99.3	99.2	99.2	99.1	99.1	99.0	98.9	98.9	98.9
5.5	97.7	97.6	97.4	97.3	97.1	96.9	96.7	96.5	96.3
6.0	93.2	92.8	92.3	92.0	91.4	90.8	90.3	89.7	89.1
6.5	81.2	80.2	79.2	78.1	77.0	75.8	74.6	73.4	72.1
7.0	57.7	56.2	54.6	53.0	51.4	49.7	48.2	46.6	45.0
7.5	30.1	28.9	27.5	26.3	25.0	23.8	22.7	21.6	20.6
8.0	12.0	11.4	10.7	10.0	9.6	9.0	8.5	8.0	7.6
8.5	4.1	3.9	3.9	3.4	3.2	3.0	2.9	2.9	2.5
9.0	1.3	1.3	1.2	1.1	1.0	1.0	0.9	0.9	0.8

Table 5 : Percentage unionized Ammonia in water at different pH values and temperature

Temp. °C									
pH	*16*	*18*	*20*	*22*	*24*	*26*	*28*	*30*	*32*
7.0	0:30	0.34	0.40	0.46	0.52	0.60	0.70	0.81	0.95
7.2	0.47	0.54	0.63	0.72	0.82	0.95	1.10	1.27	1.50
7.4	0.74	0.86	0.99	1.14	1.30	1.50	1.73	2.00	2.36
7.6	1.17	1.35	1.56	1.79	2.05	2.35	2.72	3.13	3.69
7.8	1.84	2.12	2.45	2.80	3.21	3.68	4.24	4.88	5.72
8.0	2.88	3.32	3.83	4.37	4.99	5.71	6.55	7.52	8.77
8.2	4.49	5.16	5.94	6.76	7.68	8.75	10.00	11.41	13.22
8.4	6.93	7.94	9.09	10.30	11.65	13.20	14.98	16.96	19.46
8.6	10.56	12.03	13.68	15.40	17.28	19.42	21.83	24.45	27.68
8.8	15.76	17.82	20.08	22.08	24.88	27.64	30.68	33.90	37.76
9.0	22.87	25.57	28.47	31.37	34.42	37.71	41.23	44.84	49.02
9.2	31.97	35.25	38.69	42.01	45.41	48.96	52.65	56.30	60.38
9.4	42.68	46.32	50.00	53.45	56.86	60.33	63.79	67.12	70.72
9.6	54.14	57.77	61.31	64:54	67.63	70.67	73.63	76.39	79.29
9.8	65.17	68.43	71.53	74.25	76.81	79.25	81.57	83.68	85.85
10.0	74.78	77.46	79.92	82.05	84.00	85.82	87.52	89.05	90.58
10.2	82.45	84.48	86.32	87.87	89.27	90.56	91.75	92.80	93.84

Table 6 : Concentrations of Ammonia ($NH_3 + NH_4^+$) which contain an union-ized Ammonia concentration.

pH Value *Temperature*	*6.5*	*7.0*	*7.5*	*8.0**	*8.5**	*9.0**
5	63.3	20.0	6.3	2.0	0.66	0.23
10	42.4	13.4	4.3	1.4	0.45	0.16
15	28.9	9.2	2.9	0.94	0.31	0.12
20	20.0	6.3	2.0	0.66	0.22	0.088
25	13.9	4.4	1.4	0.46	0.16	0.069
30	9.8	3.1	1.0	0.34	0.12	0.056

* Concentration of ammonia derived for pH values above 8.0 may too stringent if the concentration of free CO_2 in water is very low.

Table 7 : Specific gravity and normality of important acid and Bases.

Compound	*Approximate concentration percentage*	*Specific gravity*	*Normality of concentrated reagent*	*ml. of concentrated reagent for 1 lit of N solution*
H_2SO_4	96.0	1.84	36.0	28.0
H_3PO_4	85.5	1.65	44.4	22.5
HNO_3	70.4	1.50	15.9	63.0
Hcl	37.2	1.18	12.1	82.5
$H_4C_2O_2$	99.8	1.13	17.4	57.5
HF	49.0	0.99	28.9	34.5
NH_4OH	56.6	0.90	14.5	69.0
NaOH	50.5	2.13	19.4	51.5
KOH	45.0	2.04	11.7	85.5

Table 8 : Table for the dilution of 96% alcohol (Rectified Spirit)

Grade required (%)	*Volume of rectified spirit (ml)*	*Volume of distilled water (ml)*
90	93.5	6.5
80	83.5	16.7
70	72.9	27.2
60	62.5	37.5
50	52.1	47.9
40	41.6	58.4
30	31.2	68.8

Table 9 : Table for dilution of commercial formalin

Grade of formalin required (%)	*Volume of 40% formalin (ml)*	*Volume of distilled water to be added (ml)*
3	7.5	92.5
5	12.5	87.5
8	20.0	80.0
10	25.0	75.0
15	37.5	62.5
20	50.0	50.0

Table 10 : Friction loss of water, in feet per 100 ft of clean wrought-iron or steel pipe*

	Nominal diameter of pipe, in inches											
Flow, gpm	*1*	*1¼*	*1½*	*2*	*2½*	*3*	*4*	*5*	*6*	*8*	*10*	*12*
5	1.93	0.51										
10	8.86	1.77	0.83	0.25	0.11							
14	12.8	3.28	1.53	0.45	0.19							
20	25.1	6.34	2.94	0.87	0.36	0.13						
24	35.6	8.92	4.14	1.20	0.50	0.17						
30	54.6	13.6	6.26	1.82	0.75	0.26	0.07					
40		23.5	10.79	3.10	1.28	0.44	0.12					
50		36.0	16.4	4.67	1.94	0.66	0.18	0.06				
75			35.8	10.1	4.13	1.39	0.28	0.12				
100			62.2	17.4	8.51	2.39	0.62	0.20	0.08			
120				24.7	10.0	3.37	0.88	0.20	0.12			
150				38.0	15.4	5.14	1.32	0.33	0.17			
170				48.4	19.6	6.53	1.67	0.54	0.22			
200				66.3	26.7	8.90	2.27	0.74	0.30	0.08		
220					32.2	10.7	2.72	0.88	0.36	0.09		
260					44.5	14.7	3.24	1.20	0.49	0.13		
280					51.3	16.9	4.30	1.38	0.56	0.14		
300						19.2	4.89	1.58	0.64	0.16		
340						24.8	6.19	2.00	0.81	0.21		
400						33.9	8.47	2.72	1.09	0.28	0.09	
500						52.5	13.0	4.16	1.66	0.42	0.14	0.06
600							18.6	5.88	2.34	0.60	0.19	0.08
700							25.0	7.93	3.13	0.08	0.26	0.11
800							32.4	10.22	4.03	1.02	0.33	0.14
900							40.8	12.9	5.05	1.27	0.41	0.17
1000							50.2	15.8	6.17	1.56	0.50	0.21
1100								19.0	7.41	1.87	0.59	0.25
1200								22.5	8.76	2.20	0.70	0.30
1300									10.2	2.56	0.82	0.34
1400									11.8	2.95	0.94	0.40
1500									13.5	3.37	1.07	0.45
2000									23.8	5.86	1.84	0.78
3000										12.8	4.00	1.68
4000										22.6	6.99	2.92
5000											10.80	4.47

* Reprinted from "Tentative Standards of Hydraulic Institute, Pipe Friction," Copyright 1948 by the Hydraulic Institute. 122 E. 42d St. New York, 10017.

SELECTED READINGS

1. **Advances in Fish and Wildlife Ecology and Biology Vol. 1/***(Ed.) B. L. Kaul*
2. **Advances in Fish and Wildlife Ecology and Biology Vol. 2/***(Ed.) B. L. Kaul*
3. **Aquaculture and Risk Management/**An FAO Fishing Manual/*PAD Secretan and C. E. Nash*
4. **Aquaculture Project Formulation/**An FAO Fishing Manual/*David Insull and Colin E. Nash*
5. **Aquaculture Systems and Practices:** A Selected Review/An FAO Fishing Manual/*Elvira A. Baluyut*
6. **The Biology of Fishes/**Introduction by Dr. Vijay Dev Singh/ *Harry M. Kyle*
7. **Chemical Analysis of Fish Pond Soil and Water /** *G.N. Chattopadhyay*
8. **Conservation and Management of Aquatic Resources /** *(ed.) Ashutosh Gautam*
9. **Current and Emerging Trends in Aquaculture /** *(Ed.) P.C. Thomas*
10. **A Descriptive Catalogue of the Indian Deep-Sea Fishes in the Indian Museum/***A. Alcock*
11. **Ecological and Fisheries Development in Wetlands**—A Study of Chilka Lagoon/*K. P. Biswas*
12. **Ecological Imbalance of the Ganga River System—Its Impact on Aquaculture /** *Parmila Ray*
13. **Economics of Brackishwater Shrimp Culture /** *P. Andrew*
14. **Economics of Fisheries:** A Case Study of Andhra Pradesh/*N. Subba Rao*
15. **Economics of Fisheries Management—** A Critique in Third World Perspective/ *R. Korakandy*
16. **Farm-made Aquafeeds /** FAO
17. **Fish and Fisheries /** 2nd revised and enlarged edition / *B.N. Yadav*
18. **Fish Endocrinology/***B. N. Yadav*

19. **Fisheries Science/***R. Santhanam*
20. **Freshwater Biology /***Sir S.F. Harmer*
21. **Freshwater Ecosystem of India /** *K.Vijaykumar*
22. **The Freshwater Guide to the Aquarium /***L.I. Wills*
23. **Freshwater Prawn Farming—A Manual for the Culture of Macrobrachium Rosenbergii/** *Michael B. New and Somsuk Singholka*
24. **Geographic and Oceanographic Research in Indian Waters/***R. B. Seymour Sewell*
25. **Guidelines for the Promotion of Environmental Management of Coastal Aquaculture Development /** *U.C. Barg / FAO*
26. **Handbook of Mollusca and Radiata /** *Baron Cuvier*
27. **Handbook of the Freshwater Fishes of India /** *Captain R. Beavan*
28. **Harvesting Aquatic Resources/***K.P.Biswas*
29. **A History of Fishes /** *J.R. Norman and Tenison*
30. **Indian Marine Biology/***G. Seshappa*
31. **Limnological Research in India/***S.R. Mishra*
32. **A Manual of Freshwater Ecology:** An Aspect of Fishery Environment/*R. Santhanam et. al.*
33. **Marine Fish-Farming for India /***James Hornell*
34. **Marine Fisheries of India /** *P. Bensam*
35. **Microscopic Fresh Water Life /** *FJW Plaskitt*
36. **Oils Fats and Fatty Foods—Their Practical Examination /** *E. Richards Bolton*
37. **Pond Construction for Freshwater Fish Culture — Pond Farm Structures and Layouts**—Simple Methods for Aquaculture/*A.G. Coche* and *J.F. Muir*/*FAO*
38. **Production and Marketing Management of Marine Fisheries in India /** *R. Sathiadhas*
39. **Recent Advances in Fish Ecology, Limnology and Eco-Conservation Vol. 3/** *(Ed.) Surendra Nath*
40. **Recent Advances in Fish Ecology, Limnology and Eco-Conservation Vol. 4/** *(Ed.) Surendra Nath*
41. **Recent Researches in Aquatic Environment/** *Ashutosh Gautam & N. K. Agarwal (Eds.)*

42. **Reservoir Fisheries of India/** *FAO*
43. **Role of Fisheries in Rural Development/** *S. Giriappa* (Ed.)
44. **Technological Change and the Development of Marine Fishing Industry in India:** A Case Study of Kerala/ *Ramakrishnan Korakandy*
45. **Vaccine Manual—The Production and Quality Control of Veterinary Vaccines for use in Developing Countries /** *Noel Mowat and Mark Rweyemamu/ FAO*
46. **मीठा जल झींगा पालन:** मैक्रोब्रैकियम रोजनबर्गी पालन की एक नियमावली/ *माइकल बी. न्यू और सोमशोक सिन्होलक / एफ. ए. ओ.*
47. **मछली पलन के नए आयाम** / *अखिलेश कुमार सिंह*
48. **मीठा जल मछली पालन के लिए तालाब निर्माण** : तालाब प्रक्षेत्र रचनायें ले आउट / *एफ. ए. ओ.*